John Sterling Kingsley

The Naturalist's Assistant

A Handbook for the Collector and Student

John Sterling Kingsley

The Naturalist's Assistant
A Handbook for the Collector and Student

ISBN/EAN: 9783337025984

Printed in Europe, USA, Canada, Australia, Japan

Cover: Foto ©berggeist007 / pixelio.de

More available books at **www.hansebooks.com**

THE

NATURALIST'S ASSISTANT

A Hand-Book for the Collector and Student

WITH A BIBLIOGRAPHY OF FIFTEEN HUNDRED WORKS
NECESSARY FOR THE SYSTEMATIC
ZOÖLOGIST

BY J. S. KINGSLEY

BOSTON
S. E. CASSINO, PUBLISHER
1882

CHAPTER I.

COLLECTING AND PRESERVING SPECIMENS.

MANY treatises and papers have been written on the methods of collecting and preserving zoölogical specimens; the more important of which are enumerated below. Space prevents giving the various operations in detail in this volume, but it is hoped that the directions given, although concise, will prove explicit and valuable. For more extended accounts of the methods employed in collecting and preserving specimens, the student is referred to the following works:

Boitard — Manuel du Naturaliste Préparateur. Paris, 1853.
T. Brown — The Taxidermist's Manual. London, 1859.
Elliott Coues — Field Ornithology. Salem, 1874.
J. B. Davies — Naturalists' Guide. Edinburgh, 1853.
G. Dimmock — Directions for the Collecting of Coleoptera. Springfield, Mass., 1872.
J. H. Emerton — Life on the Seashore. Salem, 1880.
James Lewis — Directions for Collecting Land and Fresh Water Shells (American Naturalist, vol. ii, 1868).
C. J. Maynard — Naturalists' Guide. Salem, 1870.
A. S. Packard, jr. — Directions for Collecting and Preserving Insects (Smithsonian Institution).
Smithsonian Directions for Collectors.
W. Swainson — Taxidermy. London, 1851.

C. A. Walker — Hints on Taxidermy (American Naturalist, vol. iii, 1870).

Lord Walsingham — Directions for Collecting Micro-Lepidoptera (American Naturalist, vol. vi, 1872).

S. P. Woodward — Manual of the Mollusca. London, 1871.

VERTEBRATES.

Mammals and birds are most readily procured by shooting with a gun, using shot large enough to kill, but not so large as seriously to injure the specimen. The size of the shot to be employed cannot, of course, be dogmatically prescribed, as it varies with the size of the animal, but in general terms "number 8" shot will be large enough for all birds under the size of a pigeon, while for birds of greater bulk, "number 5", or larger, will be required. These remarks apply equally well to the smaller mammals; for the larger ones a rifle may be necessary. It must be insisted on that the collector shoot at any part of the body rather than at the head. Some collectors use a bow and arrow or a blow gun for the smaller birds, and with slight practice become very expert. Traps and snares of various sorts are frequently employed and with the advantage of obtaining the specimen in an uninjured condition. "Bird-lime" is also used to capture birds alive.

The English method of making this substance is as follows: the middle bark of the holly, mistletoe or distaff-thistle, is chopped up and boiled in water several hours. The resulting liquid is then strained and concentrated by evaporation until it assumes a gelatinous consistency, resembling moist putty. Doubtless the bark of several of our American trees and shrubs would answer the same purpose, but the writer is not aware of any experiments having been tried. A substitute may be made by taking ordinary wheat flour, placing it in a bag of fine muslin and washing it in running water, aiding the process by squeezing until all the starch is washed out, and only the *gluten* remains behind. This gluten is an adhesive substance, which is said to answer the purpose well.

A third formula for bird-lime is to take linseed oil and heat it over a slow fire (carefully watching it to see that it does not burn), until it is very thick, then pour it into cold water. If it should prove too thick, the addition of a little pine tar will readily thin it for use.

The bird-lime should be smeared on the branches of trees, etc., where birds most do congregate, and by adhering to their feet, it holds them fast, and renders them an easy prey to the collector.

No matter how procured, all mammals and birds intended for stuffing should have the mouth, nostrils, anus and all wounds, stopped immediately with cotton wool to prevent any soiling of the fur or feathers. It is also well to place each bird head first in a cone made of cartridge paper, before placing in the game bag, as this will prevent disarrangement of the feathers.

All Vertebrates are really more valuable as alcoholic specimens, than they are when mounted after the usual manner of taxidermists, as the naturalist is then able at any time to pursue any desired investigation of their anatomy, a course from which he is utterly debarred with stuffed specimens. Before being placed in spirit, the abdominal walls of all Vertebrates should be cut open, care being taken not to injure the viscera. This allows the alcohol to readily penetrate the interior. It is also well to remove a portion of the skull, so that the preservative fluid can have access to the brain. Alcoholic specimens of foreign vertebrates thus prepared are a great desideratum in all museums, and especially in those where it is realized that science is more than skin deep, and consists of more than a lot of scientific names.

The art of skinning mammals and birds may be more readily learned by seeing another perform the operation than from pages of description. For those who do not have an opportunity of learning the methods employed by observation, the following directions which are modified from those given in Davies' " Naturalist's Guide " (by the way a very valuable little work) may prove of use.

MAMMALS.

The cotton wool is first removed from the nostrils, mouth, anus and wounds and replaced by fresh plugs. The animal is then laid on its back, its legs pressed out and the fur parted on the median line of the ventral surface. An incision is then made through the skin, at the posterior portion of the abdomen, care being taken to cut the skin only and not the underlying muscles, this incision to be continued forward to near the neck. With the left hand the skin is then raised first on one side and then on the other, and at the same time separated from the adjacent muscles with the *handle* of the scalpel, an ivory paper knife or other blunt instrument held in the right hand. The portion of the skin thus disengaged is kept from adhering to the flesh of the body, by being sprinkled with plaster of Paris. The anus is then cut through, and immediately after, the tail at its junction with the body. The hind legs are then cut off at the upper thigh joint, and the posterior part of the body turned out of the skin. The carcase is now suspended by the pelvis on a hook supported by a string from the ceiling of the room, and the skin gently pulled down from the back, the operation being facilitated by the handle of the scalpel as before. The fore legs are then disarticulated at the shoulder joint. The neck is then uncovered and the head proceeded with. In skinning the latter part, great care must be exercised to cut off the ears as close to the skull as possible, and to preserve the eyelids, nostrils and lips uninjured. The neck is now separated from the skull. The trunk is now removed from the hook and laid aside, and the legs successively hung on the hook, and the skin drawn down as far as the toes. The flesh is then removed from the bones of the legs, care being taken to leave the tendons uniting the joints entire. In order to skin the tail, the first two or three vertebræ are laid bare and attached to a stout cord. A cleft stick is then made to embrace this portion already skinned beyond the cord and gradually forced down toward the extremity, carrying with it the freed skin.

    The skin now being separated is carefully examined and any flesh or fat removed by the scalpel. The inside of the skin is then thoroughly rubbed over with the common white arsenic of the shops (arsenious acid) or if preferred completely anointed with arsenical soap. The bones of the legs are to be treated in the same manner, and, having been wrapped with tow, are returned to their places. The skull is next pulled out through the neck and freed from fat and flesh and the brain removed through the opening behind. In some cases it may be necessary to enlarge this opening by breaking away the adjacent bone, but this course should be avoided as much as possible, as the skull, from a scientific standpoint, is ot as much value as the skin, and should the latter by any means become destroyed (by no means an uncommon occurrence), the specimen will still retain a scientific value. It is well, when possible, to remove the skull entirely from the skin and macerate it in water until the flesh is removed, and the brain so decomposed as to be readily shaken out of the opening. It is sometimes desirable to preserve the skull and the skin separately, and at such times a rough model of the skull may be made of plaster of Paris, and placed in the skin, while such disposition is made of the skull as may be desired. Should the skull be returned to the head, the place of all flesh removed should be filled by tow. A wire wrapped with tow may be inserted in the tail, while the body is distended to something like its original shape by the same material.

BIRDS.

A paper ring is made fitting tightly around the body; this is preserved as a measure of the proper size and is used farther on. This ring is then removed, the bird laid on its back, with the head pointing obliquely from the operator toward his left hand. The feathers are then separated in the median line by the left hand, and an incision is made much as in mammals, the extent of this slit varying somewhat with the expertness of the operator, as well as the kind of bird being skinned. The slit being made, the fingers are inserted between the skin and the flesh, and the parts exposed dusted with plaster of Paris, to prevent any adhesion of the feathers. In some cases, it is advantageous to sew strips of cloth to the cut edges of the skin to keep the feathers clean, and also to prevent the skin from stretching. The legs are now pushed forward, and divided at the knee joints, after which the vertebral column is divided, leaving the last joint in the skin, as a support to the tail feathers. The body is then suspended from the hook by the rump end, and the skin separated from the back and sides (as in the case of mammals) until the shoulder joints appear. If the bird in hand be a water-fowl, it may be necessary to separate the wings at the shoulder joint, but whenever possible the division had best be made at the elbow. The neck is next to be skinned, taking great care not to stretch the skin, especially in the case of the long-necked birds. Then the head is separated from the integument as far as the bill. Now remove the tongue and muscles from the skull, and separate it from the neck, placing the carcase aside, and remove the brain from the skull with a quill, enlarging the opening if necessary for the purpose. Great care should be taken, in skinning the head, not to injure the external ear and the parts around the eyes.

The bones left in the legs (*tibiæ*) are now to be skinned, cleaned, thoroughly covered with preservative (arsenic or arsenical soap), and wrapped with tow. After treating the skin of the leg with arsenic, the bones are to be returned to their places by being gently pushed in. When the upper bone of the wing (*humerus*) is retained, it must be treated in the same manner. Except in the case of large birds, no treatment is necessary for the bones of the fore wing. In these, however, the muscles may be removed by making an incision on the inside of the wing, and then impregnating with arsenic, and fastening with two or three stitches. Now remove all flesh and fat from the skull and skin, and impregnate them thoroughly (the skull inside as well as out) with arsenic. A wire about the length of the neck is then taken, and one end being fastened in the base of the skull, a little tow or flax filled with arsenic is wrapped around it, and the head is pulled out of the neck by means of a string attached to the bill, bringing with it the tow-covered wire. Next dispose the wings in their proper position, place the paper ring, mentioned above, around the body, stuff the skin out to its proper dimensions with tow, sew up the slit, label and dry, and the specimen is ready for the cabinet.

The foregoing directions are applicable to the majority of birds but will have to be modified occasionally. The feet of the larger birds of prey are frequently fleshy. In these cases it will be necessary to cut a slit on the under side of each toe and perhaps up to the back of the tarsus to remove the muscles and tendons; then rub in the preservative, fill with tow and close the openings with a few stitches. The webs on the feet of swimming birds had best be skinned below and in all cases should be thoroughly poisoned.

When the head is of such a size that the skin of the neck cannot be drawn over it, as is the case with the flamingo and most web-footed birds, it will be necessary to make an incision in the neck near the base of the skull and through it remove the brain, etc. This is an operation of considerable nicety as the feathers are very liable to get daubed. In case, however, any blood, brain or feculent matter should get on the feathers, it should be carefully removed by a cloth *dampened* in warm water. Grebes and other water fowl with white silky bellies are sometimes skinned from an incision in the back. In this way the feathers are less liable to be daubed and to be stained by the oil of the body when in the cabinet.

Humming birds from their diminutive size are not easily skinned. They may, however, be preserved by making an incision on the belly and removing as much of the soft parts as possible with the forceps and scissors. The skin should then be thoroughly poisoned and filled with cotton wool or tow.

On the label attached to each bird should be information as to the following points :

Exact locality, date of capture, sex, food (ascertained by an examination of crop and gizzard) color of the eyes, feet, bill, gums, membranes, caruncles, etc. Attitude of body when at rest. Does the bird perch or not? The length in inches from the tip of bill to the end of tail, the distance between the extremities of the outstretched wings and the length of the wing from the carpal joint.

Should it be desired to mount the specimen, information on the following points will aid the taxidermist in giving the proper position.

Position of the wings whether supported or hanging, crossing on the tail or not. Are they continuous, or covered by the feathers of the back and breast, for the upper half or third or two-thirds of their length? Do their extremities reach the tip of the tail, the half or fourth of its length? Are the heels covered by the feathers of the belly?

The skins of mammals and birds prepared according to the foregoing directions are really more valuable for the naturalist than the mounted specimens. They may be kept systematically arranged in boxes or drawers. Mounting mammals and birds is the work of a taxidermist and directions for the operations are foreign to the purposes of this work. If it be desired to prepare the specimens for exhibition they had better be sent to the professional taxidermist, as amateur work generally presents a very slovenly appearance. One thing, however, should always be insisted on; the stands employed should be of the simplest character. For birds the form of stand shown in the adjacent figure is preferable. These stands are usually painted white. For mammals and many aquatic birds a board is all that is necessary. By all means avoid the use of moss, mica sand, artificial leaves, etc., as they not only afford excellent lurking places for vermin, but also detract greatly from a scientific appearance of the collection; they and not the specimens attract the eye.

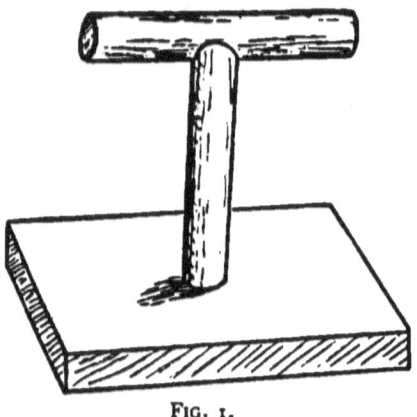

FIG. 1.

NESTS AND EGGS.

The nests and eggs of birds are largely collected, especially by the young, and many a naturalist of note traces his interest in zoölogy to his early cabinet of eggs. It is not necessary to tell where to look for nests as every one with his eyes open can find them. Some are in tall trees, some birds build in bushes, some on the ground, while others affect the habitation of man.

When a nest is found, great care should be taken to identify the bird to which it belongs, otherwise both nest and eggs are scientifically useless. In some cases and especially with collectors in foreign countries, it will be necessary to shoot the parent birds to accomplish this. All of the eggs should be taken and labelled at once so as to be beyond a doubt at any time. The contents should also be extracted. This is effected by boring a hole in *one side* of the egg with an "egg drill" (a steel instrument which can be procured of any dealer in naturalists' supplies), fig. 2, and through this opening all contents may be withdrawn. For this purpose some use a fine nozzled syringe, while others insert the tip of a jeweller's blow-pipe into the opening, and then by blowing in the egg the yolk is forced out around the sides of the pipe. If incubation has proceeded to any length it will be necessary to cut the embryo up with fine pointed scissors and extract it in fragments with the aid of a bent needle. The interior should then be thoroughly rinsed, first with water and then with

FIG. 2.

arsenical soap. The opening can then be covered with a bit of goldbeater's skin.

Exact labels giving the name of the bird laying the eggs, the locality and date, the number of eggs in the nest, etc., should be kept with each specimen and numbered to correspond with a number placed on the egg.

Eggs of our native birds taken at various stages of incubation, the shell cracked and then the whole placed in 40 per cent. alcohol and in a few hours transferred to fresh and slightly stronger spirit, and then after a day or two to alcohol of 70 to 80 per cent., would be very desirable in every museum of the world. Of course, with each egg should be preserved not only the name of the species, but also the number of hours since incubation began. Thus the student will be able to trace more or less completely, according to the amount of material at command, the development of the various forms of which, at present, comparatively little is known. The same process may be advantageously followed with the eggs of other animals, and in dissecting mammals all embryos should be carefully preserved.

REPTILES AND BATRACHIA.

The collection of snakes, turtles, lizards, frogs, toads, and salamanders is not accompanied with any special difficulty though proper precautions should be taken against venomous serpents. Various species of each group affect certain localities, some living on land and others in the water. Some live in the open fields, others in thick woods, while still others are generally found in damp places under decaying timber, etc.

All of the lower vertebrates are best preserved as "wet specimens," and in fact with the exception of the turtles and a few large forms are spoiled by being skinned and stuffed. In skinning turtles the lower shell (plastron) should first be removed with a chisel or saw; the succeeding steps are essentially the same as pursued with mammals. Alligators and large lizards are skinned the same as mammals. When it is desired to put any of the lower vertebrates in alcohol, an incision should be made in the abdominal walls, so that the spirit may more readily penetrate the viscera. This is absolutely necessary if it be desired at any future time to investigate any more of the anatomy than the osteology.

FISHES.

Besides the familiar hook and line, fishes may be obtained by seines, trawls, etc., to be described further on under the head, "Marine Collecting." A good way of obtaining many forms is to visit the fish markets; and also if possible hire the fishermen themselves to bring in specimens of all sorts that come up in their nets or on their lines. In this way many varieties may be obtained which never appear in the markets, as fishermen are accustomed to throw back all fish which according to their ideas are not edible.

Fishes are almost universally preserved in alcohol, though some of the largest ones are occasionally stuffed. At such times a professional taxidermist had best be employed.

In putting in alcohol the abdominal walls should be opened so that the spirit may the more readily enter and thus ensure the preservation of the viscera, some parts of which are very important even from a systematic standpoint.

Fishes in alcohol do not present a very interesting or attractive appearance on the shelves of a museum, and only the ichthyologist is able to decide on the identity of alcoholic and fresh specimens. Many attempts have been made to preserve fish dry but the majority of methods employed do not produce very satisfactory results. The best process known to the writer is that invented by Dr. H. E. Davidson, who has not only described his method but has also given chances to witness the operation which is as follows:

The necessary materials are thin pieces of soft wood about one-eighth of an inch in thickness; square sticks measuring from three-fourths of an inch upwards; plaster of paris, glycerine, tissue paper, pins, and double pointed carpet tacks.

The outline of the fish without the fins is marked on two pieces of board which are held together by pieces of the square sticks tacked across the ends, and then the portion corresponding to the body is cut away so that we have two strips of wood one following the dorsal and the other the ventral contour of the fish. The fish is then placed in this opening and the various fins are extended and fixed in position with pins, the board in the meantime being supported so that one side of the fish can freely extend through the opening in the joined boards. Strips of tissue paper wet with glycerine are then laid smoothly over the fish and next a coating of plaster is poured over the same side. When the plaster is hardened, the boards, etc., are reversed and the rest of the work is carried on from the opposite side of the body. All that portion of the fish which projects through the opening is first cut away, and then all of the muscles, bones and viscera, are carefully removed until nothing remains but the skin supporting the fins and its plaster backing. In this condition one side of the skin is entire and on the other side a narrow strip of skin extends around the median line of the body from a quarter to half of an inch in width. The interior of the skin is now dusted with arsenic. The eye is then placed in position and the skin is filled with plaster mixed to about the consistency of cream. The double pointed carpet tacks are then taken, and their points, having been bent as shown in the adjacent figure, are hooked into the strip of skin and the loop embedded in the plaster. A small strip of wood (previously coated with shellac to prevent undue expansion from the moisture), is also embedded in the plaster, its upper surface being even with that of the plaster. The two halves of the board are separated when the plaster becomes dry, the skin with its plaster interior is removed from its mould and washed and the fins placed in clips so that they may dry flat. When thoroughly dry, the specimen is mounted on a wooden tablet by screws passing into the embedded block and the whole is ready for exhibition.

FIG 3

No means have yet been found of preserving the natural colors of the fish; and the only way of representing them on the specimens thus mounted is by means of paints.

This process which has been thus briefly described, is the property of Dr. H. E. Davidson of Boston, and to him all inquiries, as to the rights to use it, should be addressed.

SKELETONS.

Of fully as much importance as skins, and scarcely more difficult to prepare, are skeletons of vertebrates, and when from any circumstance it is impossible to prepare the whole skeleton, the skull can frequently be preserved. The *modus operandi* is essentially the same for all vertebrates.

Skeletons are of two sorts, natural and artificial : *i. e.*, those where the bones are united by the ligaments, and those in which the ligaments are removed as well as the flesh, and the bones are articulated with wires and rods. Natural skeletons can only be prepared when the subject is of small size ; not exceeding the fox or goose in bulk. Skeletons of larger animals must be, to a greater or less extent, artificially articulated.

The skin is first removed from all parts of the body, the head separated and the viscera extracted. Then as much of the flesh as possible is removed with the scalpel, great care being taken not to cut, scratch or otherwise injure the bones. The body is then placed in cold water to macerate, sometimes a little caustic potash is added to the water to accelerate the decomposition of the flesh, but except a gain in time there are no advantages to be gained by the addition. For the first few days the water should be changed every day, and when the flesh is partially decomposed as much as possible is to be removed, taking care, if the skeleton is to be a natural one, not to injure the ligaments. The partially cleaned skeleton is then returned to the macerating tub, and on succeeding days is subjected to the cleansing operation until all the flesh is removed. It may be well, as a final step, to use a stiff nail brush to remove the last traces of flesh.

The skull is treated in the same manner, and the brain is broken up and removed with a stick, through the occipital foramen. It is sometimes desirable to open the skull by sawing off the top, and thus to remove the brain more carefully, preserving the *tentorium* and *falx cerebri* uninjured.

Under no circumstances should the bones be boiled as that operation greases them and gives the skeleton an unsightly appearance. If the water is left too long without changing, the bones are apt to become discolored.

When finally cleansed, the skeleton (if a natural one) has a wire passed down the spinal canal, its end projecting from the neck and then, being supported in the desired position by strings or wires attached to a suitable framework, is left to dry. When dry, the skull is fastened to its place on the wire projecting from the neck, by means of copper or brass wire, the lower jaw is articulated to the skull, and held in a proper position by spiral springs. The body is then supported on a couple of upright standards, arising from a horizontal base, and after being duly and fully labelled, the specimen is complete.

It would be impossible, without occupying much more space than is allowed, to describe the method of articulating an artificial skeleton, while on the other hand it can be readily understood, after a few minutes' study of one thus prepared; and therefore all who wish to articulate artificial skeletons are respectfully requested to obtain the requisite knowledge by observation. Great care, however, should be exercised that none of the small bones be lost in the process of maceration.

COLLECTING INSECTS.

Insects are the most numerous both in individuals and in species of any group of the animal kingdom and may be found almost everywhere and at every season of the year. Their beauty, their numbers, and the ease with which they may be collected and preserved, render them great favorites. Many a naturalist, who has acquired prominence, traces his studies to the collections of insects made in his youth.

FIG. 4.

The insect collector needs certain pieces of apparatus none of them expensive and all easily made by one possessing an ordinary amount of mechanical skill, or they may be readily procured in the shops. Those most essential are insect nets, means of killing, and conveniences for carrying the specimens home.

An insect net, fig. 4, is readily made by taking a stout brass wire (iron rusts too readily) and bending it into a ring about twelve or fifteen inches in diameter. The ends of the wire should be bent out and soldered into a ferrule which will fit on the end of a cane or other handle. The net proper should be about twenty inches in depth and made of gauze or mosquito netting. It should not be attached directly to the ring, as it would then quickly fray out, but to a piece of strong cotton cloth which in turn is sewed to the ring. Other more elaborate forms are made in which the ring will fold up for convenience in carrying, but the saving hardly repays the additional expense. Those interested will find a good description and figure in Dr. Packard's "Directions for Collecting and Preserving Insects," published by the Smithsonian Institution, page 4, fig. 2, where one or two other forms are also described.

Fig. 5.

The net is used principally for collecting the strong flying insects (*e. g.*, Butterflies), either on the wing or while at rest. With a swoop the net is brought over the insect, and then, by a dexterous twist, easily acquired but not readily described, the bag is thrown over the ring and the specimen is securely imprisoned. The insect may then either be pinned while in the net or transferred to the cyanide bottle to be described farther on. Lepidoptera may be killed while in the net by giving the thorax a severe pinch, of course taking care that the wings are not injured.

A shallow scoop net, fig. 5, made in the same manner as the insect net above described is useful for collecting aquatic insects. It is not necessary to detail the method of using it, as any one will readily find out for himself.

For collecting stinging insects a pair of forceps, fig. 6, made of wire, the distal extremities of which are bent into broad blades covered with netting, will prove very convenient, especially as there is no danger of being stung. The bee or other insects are caught between the blades while resting on a flower, and while a prisoner is pinned; and then, the blades being opened, the pin is readily drawn from the meshes of the netting.

An umbrella is indispensable in collecting certain forms of insects. It is held spread open in an inverted position beneath the branches of some tree or shrub, then the foliage is beaten with a stick, and the insects drop and are caught. This is especially valuable for collecting certain Coleoptera, Spiders, Microlepidoptera, Psocidæ, etc.

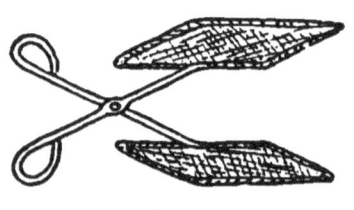

FIG. 6.

For killing insects the most convenient plan is to take a wide-mouthed bottle and place in the bottom a piece of cyanide of potassium (a dangerous poison). Then some plaster of Paris is mixed to a stiff paste with water and poured over the cyanide. The plaster soon sets and holds the chemical firmly, while its porosity allows the fumes of prussic acid to pass readily into the vacant portion of the bottle. The bottle after thus being prepared should be allowed to stand open for a day to allow the moisture from the plaster to escape; it should then be kept securely corked. One of these cyanide bottles will answer for a season's collecting and is safe for children to use. An older person, however, should prepare the bottle, as the cyanide is very poisonous and sometimes produces severe ulcers on the parts of the body with which it may come in contact.

Some instead of cyanide use ether, chloroform, benzine, or bisulphide of carbon in the bottle, but the rapidity with which these evaporate renders them far less convenient than the cyanide. Dr. Loew recommends moistening the bottom of the collecting bottle with creosote for killing Diptera. Lepidoptera may be killed by giving a severe pinch to the sides of the thorax, though this is very apt to remove many of the "feathers" from the body. The wings of a butterfly should never be touched with the fingers and great care should be taken to avoid mutilation of any insect.

For carrying specimens home the collector should be provided with wide-mouthed vials and bottles; some empty and some containing alcohol; a supply of "pill boxes" and a cork-lined box two inches in depth and in its other dimensions as large as can be conveniently carried in the pocket. Insect pins of various sizes are indispensable. The insects on being collected may be carried home alive by placing them in the pill boxes or the empty vials; or they may be killed by the cyanide bottle or being placed in the alcohol or by pinching. Beetles and bugs may be kept in the alcohol, or with other forms pinned in the field and kept in the cork-lined box. The writer has found a stiff round crowned hat a very convenient substitute for the cork-lined box, as the insects may be pinned on the inside and thus are not seen by that class of society who think a naturalist a little "cracked." Lepidoptera may be conveniently carried by folding the wings together and placing them in square sheets of paper folded into a triangular form.

It is impossible to say exactly where insects may be found. In general terms, gardens, the edges of woods and banks of ponds and streams are more bountifully supplied than treeless meadows or deep forests. In winter the moss and bark on trees cover many beetles, spiders, *Tingids* and hymenopterous insects, as well as pupæ of these and other orders. In the summer, insects are far more numerous. The open fields will afford numerous Lepidoptera, beds of flowers will attract all orders, certain forms affect mushrooms and toadstools, and *Silphidæ*, *Nitidulidæ*, and *Staphylinidæ*, as well as various flies, may be found in the vicinity of carrion. Old boards and logs afford hiding places for various larvæ as well as spiders, myriapods and beetles, while in such places the *Thysanura* thrive. In the moist loose earth at the edges of woods *Campodea*, *Trichopetalum*, *Scolopendrella* and the *Pauropidæ* should be sought. Other species of insects, notably certain *Scarabæidæ* and dipterous larvæ, live in excrementitious matter. Ponds and streams contain large numbers of insects; beetles, bugs and the larvæ of several other groups. One may do much for science by studying the transformations of these aquatic forms. Of the various stages passed through by our species of dragon-flies, caddis-flies, may-flies, etc., almost nothing is known. The galls found on trees and plants may be taken home and the larvæ contained in them reared, and the same course may be pursued with all the larvæ and pupæ found while collecting.

PINNING INSECTS.

Insects are usually mounted for the cabinet on pins made especially for the purpose, which can be procured of any dealer in naturalists' supplies. Those most generally employed are brass, silver plated. A good quality should be used, as with poor pins the specimen is apt to be covered with verdigris while the pin itself is soon destroyed. To avoid this, varnished pins, and silver and platinum wire have been employed. A gold plated pin has been recently introduced with very satisfactory results while the increase in price is slight. The best silvered pins are those made in Berlin by Klager. There are five sizes, of which numbers one, three, and five are the most convenient, number one being the finest. Still smaller pins are made for minute insects. The insect is impaled with one of these smallest pins and fastened to a bit of cork which in turn is mounted on a larger pin and the whole placed in the collection. Most insects are pinned through the thorax, but beetles should have the pin inserted through the right wing cover. The specimens should all be pinned at an equal height, so that about one-fourth of the pin extends above the insect. On the pin below the insect should be kept labels, dates and localities of capture, and all information of value. Very minute insects are frequently glued to bits of card and these are in turn pinned. It is most convenient to cover a piece of card with gum, place the insects promiscuously upon it and then when dry cut to suit the specimen. Thin pieces of mica are also used in a similar manner.

To place the insects in the cabinet, what are known as pinning forceps are frequently used. These are forceps made after the usual manner, except that the extremities are bent as shown in fig. 7, and the corrugations of the points are so arranged as to hold the pin firmly. The pin is grasped by them about a quarter of an inch from the extremity and forced into the bottom of the case with a gentle pressure. By this method all danger of bending the pins is averted, a result which frequently follows an attempt to set them with the fingers. They may also be set with much greater regularity with the forceps than without.

FIG. 7.

SPREADING BUTTERFLIES.

Butterflies and moths should always have the wings extended and it is frequently desirable to mount other insects in the same manner. This is accomplished by means of a "setting board." A strip of pine or other soft wood has a groove ploughed through the middle to the depth of from three-quarters of an inch to an inch. The bottom of the groove is generally lined with cork to hold the point of the pin. It is frequently desirable to have the surface of the setting board slightly bevelled towards the middle groove, as in this way a drooping appearance of the wings is prevented. See fig. 8.

The pin is passed through the thorax of the insect into the

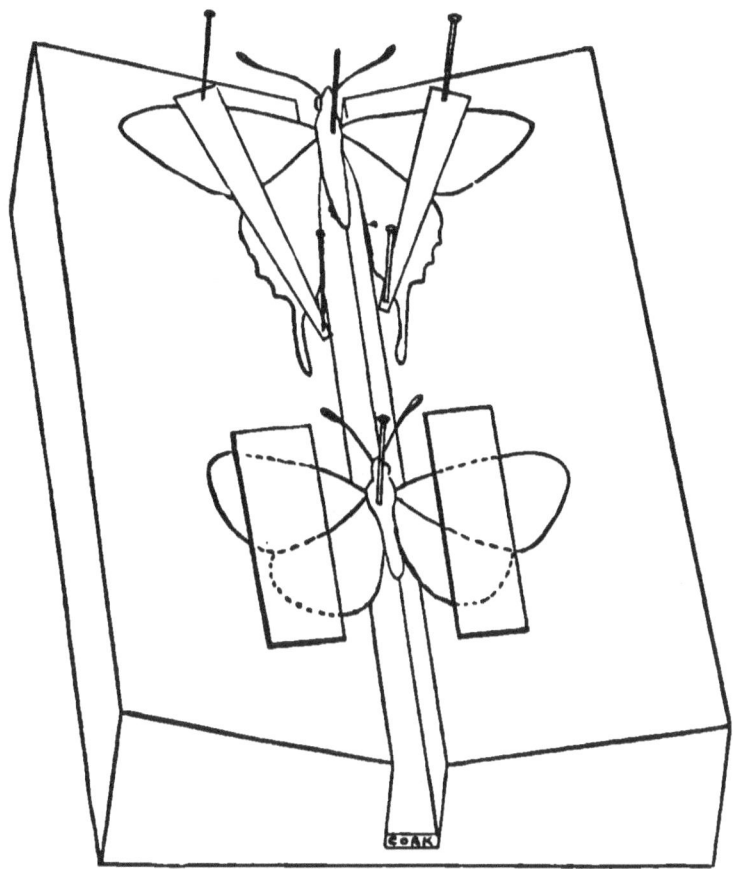

FIG. 8.

cork in the groove and then the body lying in the groove, the wings are taken, first on one side and then on the other, with a fine pair of forceps (never with the fingers), placed in the desired position and held either by a slip of cardboard pinned to the setting board, or by the weight of a piece of glass or sheet lead. The wings of the two sides should be brought to symmetrical positions and the antennæ and legs disposed as desired and held in position with pins if necessary. The board and its contents should then be put away to dry, an operation which will occupy from three days to a week or more according to the temperature and moisture of the air.

The collector in the field will find it convenient to pack butterflies in sheets of paper folded in triangles and in this shape they may be kept as long as desired, of course all due precautions being taken to prevent the ravages of injurious insects. When it is desired to spread them they may be relaxed, no matter how long a time has elapsed since their collection, by placing them in a covered box, the bottom of which is covered with moist sand, the insects being protected from soiling by the intervention of a sheet of paper. The box with its contents is then placed in a warm place for half a day when the moisture arising from the sand will have so relaxed the muscles and tendons that the specimens may be spread in the manner above described.

Beetles, bugs, and orthoptera may be taken from the alcohol and pinned at any time, but flies, Hymenoptera, Neuroptera and Lepidoptera never present a good appearance after immersion in spirits, though Hymenoptera and Neuroptera stand the operation better than the others. A protracted stay in the spirit injures the colors of all insects.

Insects frequently become broken while in the cabinet and it is recommended that inspissated ox gall made into a thick gum with a little water be employed in mending them.

It is occasionally necessary to transport collections of insects from one place to another and at such times the greatest care should be taken to protect them from injury. The collector in distant parts can send all but the Lepidoptera and flies in spirit; the former may be sent folded in envelopes while flies can only be pinned. In case mounted specimens are to be sent the danger of damage is much greater. Small cork-lined boxes should be employed and the pins should be very firmly fixed, the points being forced into the wood of the box. These smaller boxes should then be placed in a larger one and surrounded on all sides with crumpled paper, hay, "excelsior," or other elastic packing. By this process all jars received in transit are much lessened.

INFLATING LARVÆ.

Besides the usual manner of preserving larvæ in spirit they are sometimes inflated and dried. Several advantages accrue from this method of preservation; the colors are better preserved, all hairs and spines retain their proper position and the specimens are always in good condition for the artist's pencil. The following account of the process is condensed from that of Mr. Scudder:

The necessary instruments for the operation are a small

tin oven, a spirit lamp, a pair of finely pointed scissors, a little fine wire and a straw.

The oven is a tin box two and one-half inches high, two and one-half wide and five long, the cover is of glass and one end is perforated with a hole one and one-half inches in diameter. See fig. 9. No solder should be used in its construction.

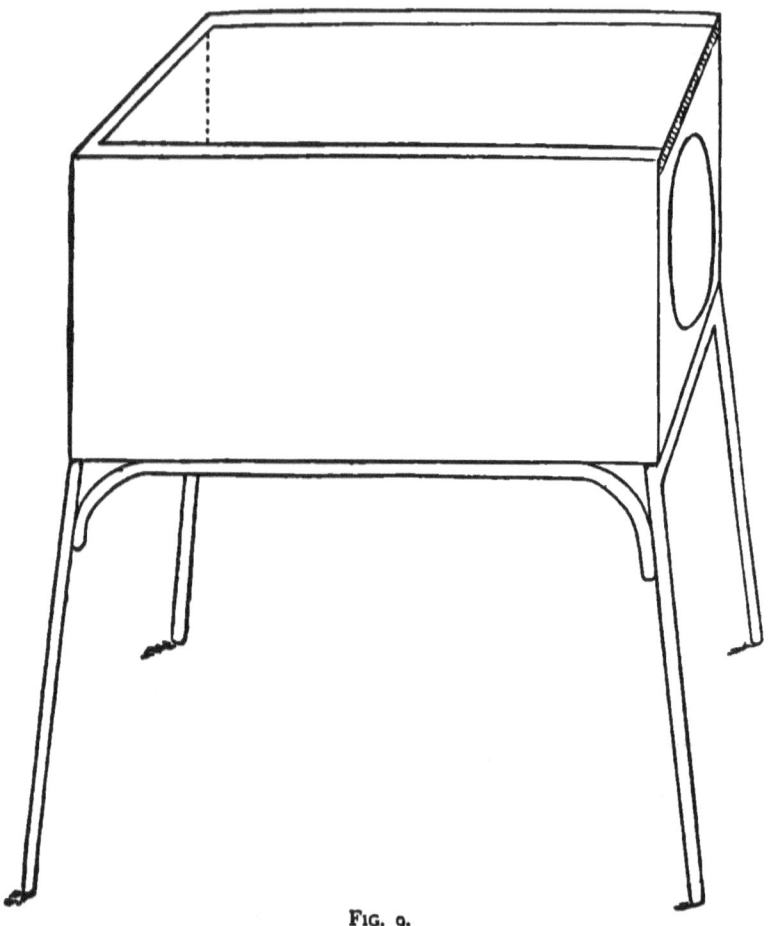

FIG. 9.

The wire should be very fine and annealed. It should not be over one-half a millimetre (one-fiftieth of an inch) in diameter.

Kill the subject by a drop of ether or by immersion in alcohol. If it be a hairy caterpillar it should remain at least half an hour in alcohol and then rest two hours on blotting paper, otherwise the hairs are apt to drop off. Then, holding the larva in the left hand, enlarge the vent slightly below with a vertical cut of the scissors, and next press the extremity of the body with the fingers so as to force out the contents of the rectum; repeat the operation a little farther forward, and so on, a slight additional portion of the contents of the body being gently pressed out each time. Great care should be exercised not to abrade the skin or get the exterior soiled by the expelled portions. When a portion of the intestine is extended, it should be seized with the forceps and, holding the head in the left hand, the tube should be forcibly but steadily torn from its attachments bringing with it most of the contents of the body.

The lamp is now lighted and placed under the oven; and a straw taken of a proper size to enter the enlarged vent; having cut off one end diagonally it is moistened and carefully inserted into the opening for about a quarter of an inch, and then pinned through the straw and anal plate with a delicate insect pin. The caterpillar is then inflated with the breath, taking care not to use too great a pressure, and then extended horizontally in the oven, the inflation being constantly continued. The posterior end should be dried first (by keeping it in the hottest portion) and gradually working forward, lastly the head. When all is dry the skin should be removed from the straw by careful use of some blunt instrument or the finger nail.

A piece of wire is then taken, a little over twice the length of the larva, and bent into the form shown in fig. 10, the free ends being slightly incurved. A drop of shellac dissolved in alcohol is then placed on the loop and the free ends are gently inserted into the body until the hinder extremity has passed half-way over the loop and the shellac has smeared the inside sufficiently to hold the specimen when dry. The folded end is then firmly wound around an insect pin and the whole, after labelling, is placed in a position where it can dry a couple of days before removal to the collection.

FIG. 10.

MOUNTING SPIDERS.

Spiders are usually preserved as wet preparations, as when dried as insects usually are, the abdomen shrinks badly. This, however, can be avoided as follows:

Kill the spider by exposure to some poisonous vapor or gas (ether, chloroform or prussic acid) and then cut the body in two between the cephalothorax and abdomen. An insect pin is then taken, its head inserted into the abdomen and its point into a stick of wood, and then the abdomen is dried by placing in the oven mentioned above, or in a test tube heated over a spirit lamp. The specimen should be kept turning so as to dry all sides evenly. When dry, the pin is cut off a short distance from the abdomen and the anterior portion of the spider is impaled on the extending part of the pin, and then a second pin being passed through the thorax (to be used in mounting in the collection), the whole is returned to the oven until dry; it is then labelled and placed in the cabinet.

BREEDING LARVÆ.

The finest specimens of Lepidoptera are obtained by rearing from the larval or pupal stages. This is accomplished by the use of breeding cages. For this purpose glass tumblers covered with gauze may be employed, but a better thing

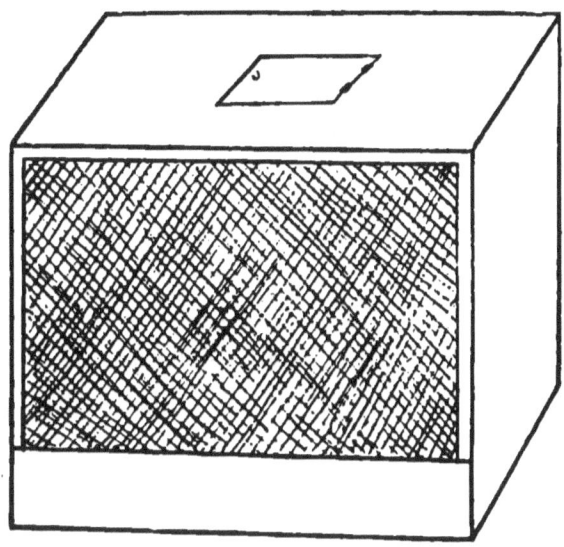

FIG. 11.

is a box especially prepared for the purpose. Take a wooden box without a cover, lay it on its side and tack a lath across the lower edge so that a shallow tray is made; then cover the rest of the opening with gauze or mosquito netting. Then put a small door in the side, which, by the inversion of the box, becomes the top. The box should then have a layer of moist earth reaching to the top of the lath. It is frequently well to subdivide the box as larvæ are occasionally apt to desert a vegetable diet and eat their companions.

Having prepared the box the larvæ should be collected and placed in it. With each larva should he collected portions of the plant on which it feeds and these should be placed in the box, inserting the ends of the twigs in the earth. When they show signs of wilting new leaves should be collected and the old ones removed.

Drawings should be made of each species of larvæ at its various stages, and in the notes which every naturalist should keep, should be noted the food plant of the larva, the dates of pupation and of the emergence of the imago, and every other item which will serve to elucidate the life history ot the insect. Frequently it is well, when a female insect has been obtained, to attempt to rear others from the egg. The insect should be furnished with that kind of food most relished by the larva and allowed to deposit its eggs on it. The date of oviposition, the size and shape of the eggs with their markings and ornamentation and the date of hatching, should be carefully preserved by means of notes and drawings. Many of the coleopterous and dipterous larvæ are carnivorous and should be supplied with meat. Other larvæ are aquatic and these must be reared in aquaria, over the top of which gauze or musquito netting has been stretched. The breeding cage should be kept in a light, airy position but should not be exposed to the direct rays of the sun. The earth in the bottom should be kept moist, otherwise the health of the larva is endangered.

Most larvæ enter the pupa stage in the fall, some climbing up the wall of the cage and spinning a cocoon, while others burrow in the earth and there pass the chrysalis portion of their life. At the approach of cold weather, the breeding cage should be removed to the cellar and kept there until spring. The collector in his trips through the woods and fields will find many pupæ; these should be brought home and placed in the breeding cage and the imago obtained. This method of breeding insects in confinement has many advantages, the most prominent being that the imagos obtained are perfect and not in that rubbed condition which is frequent in those caught with the net.

Occasionally, a larva will fail to go through its proper changes. This is generally caused by the presence of some parasite. The most common of these parasites are Ichneumon larvæ. The adult ichneumon stings the larva and lays its eggs; these hatch and the progeny live on the juices and tissues of its host until at last it succumbs, and then the parasites go through their changes and finally emerge as perfect insects. These ichneumon flies should be carefully preserved with full notes of the host, etc.

Spiders and Myriapods may be found everywhere and are best preserved in spirit. With spiders should be preserved careful notes of colors, and the form of the web, whether vertical or horizontal, flat or dome shaped, etc. Especial pains should be taken to collect the male which is much smaller than the female and is frequently found with it. The two cannot be kept together alive as the female is so fond of her mate that she frequently eats him. Myriapods are rather difficult to preserve, because their integument is so thick that the alcohol does not readily penetrate and therefore the tissues of the body decay and the specimen falls to pieces.

If a few of the rings be punctured so as to admit the spirit to the interior of the body this may be prevented. The spirit should also be frequently changed during the first few days.

Along with the myriapods will generally be found the terrestrial crustacea (*Oniscidæ*), known under the common names of "Sow-bugs" and "Pill-bugs." These should also be carefuly collected and preserved. These forms are greatly desired as they have been almost wholly neglected by American naturalists and but little is known of our native species. Many of them, however, seem to be identical with those of Europe, and no one should attempt to describe them without access to the works of Brandt, Lereboullet, Kinahan, etc.

MARINE COLLECTING.

Every portion of the sea teems with life, which varies, not only according to geographical position but also with depth, character of bottom, temperature of water, etc., etc. In different circumstances different methods are employed for collecting.

Certain forms, principally Amphipod crustacea and shells, may be found on the shore much above high-water mark. Twice during the day the receding tide leaves a portion of the shore uncovered, and this portion "between tide marks" has its peculiar fauna. For collecting here, one should wear rubber boots and be equipped with bottles, forceps, etc. There should be a considerable variety in the bottles employed. A good idea is to have four or six large jars with wide mouths kept upright in a basket; one jar may be partly filled with alcohol and corked, the others are to contain salt water, and should remain open. In the pockets of the collector should be carried a large number of "homœopathic vials," some empty and others with alcohol. A spade and a long-handled net will also prove useful. Many forms will be found in the seaweed covering the rocks; the rocks themselves should be closely examined, turning them over for the purpose if possible, and the mud of the shore should be turned up with the spade for worms, shells, etc. The speci-

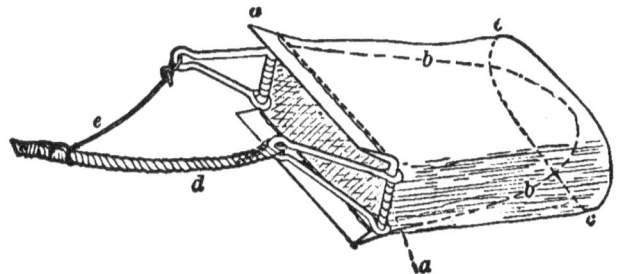

FIG. 12.

mens should be placed in one of the jars of water unless it be thought best to place certain forms at once in alcohol.

For collecting the forms from that portion which is never uncovered other methods have to be employed. Dredging is the principal one, others are the use of the trawl, the "tangle," "pumping," etc.

The dredge, fig. 12 is a rectangular frame of iron with two scrapers so that it will work no matter which side up it may fall. To this frame a net is laced by means of copper wire passed through holes in the scrapers and around the frame at the end. This net should be surrounded with a stout canvas or leather cover, so that it will not catch and tear on rocks, etc., on the bottom. Iron handles are attached to the frame, and to one of these the dredge rope is tied; the other handle is made fast to the rope by means of a bit of "spun yarn" or lighter cord, as shown in fig. 13. The object of this is that, if the dredge be caught on a rock, the smaller cord will break and the dredge free itself. A weight should be attached to the rope a short distance (six to twelve feet) in front of the dredge, so that its mouth may be kept on the bottom. The length of rope paid out should be about twice the depth of the water in which the dredging is conducted. In dredging from a sail-boat it is best, if possible, to take advantage of the currents. Put the dredge over the bows, taking care that it does not turn inside out

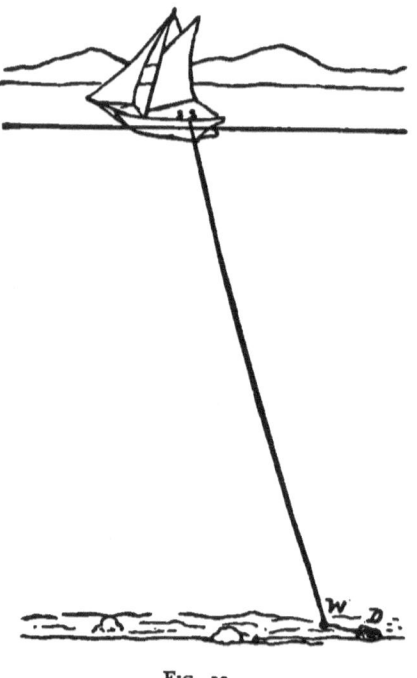

Fig. 13.

or foul while going down, then let the boat drift as though she were dragging an anchor. Where currents are not strong enough the boat head should be turned toward the wind, so that it will just move through the water, and then put the dredge over the windward side near the stern. The boat will then drift slowly, drawing the dredge over the bottom.

It is advisable to have the end of the dredge rope fastened by a "safety cord," that is, a small cord tied around the dredge rope and made fast to a cleat. Thus, in case the dredge suddenly catch, the strain will break the smaller cord and allow time for the boat to be turned around without breaking the rope and loosing the dredge. When dredging from a large boat in deep water this is absolutely indispensable, if any regard be had for the preservation of the collecting apparatus. An experienced person can judge of the character of the bottom, the condition of the dredge, etc., by the tremor of the dredge rope.

When full, the dredge should be pulled up, its contents poured into sieves and then washed with water. The sieves employed should be made of copper wire and have fine meshes. They may be so arranged as to hang over the side of the vessel, or they may be placed in a trough which will carry away the dirty water without soiling the boat. When washed, the contents of the dredge are picked over and the specimens preserved according to their character. While this operation is in progress, the dredge may be down gathering new treasures.

The trawl, fig. 14, generally consists of a long beam, six to ten feet in length, bearing a runner on each end, and attached to the beam is a long net whose lower edge is weighted with lead. This net should have several "pockets" and the hinder end should have an opening, secured with a string. The trawl is attached to the rope and used in a similar manner to the dredge. It can only be used on smooth bottoms free from rocks, and catches the fishes, shrimps, hydroids, etc., which affect such places. Sometimes, instead of having a "beam," the trawl has two "wings" made of wood and loaded on one edge so that they maintain an upright position. These wings are attached to the rope after the fashion of a kite so that the passage through the water forces them widely apart. In this form, the upper edge of the net should be floated with cork.

The tangle, fig. 15, is another piece of apparatus for

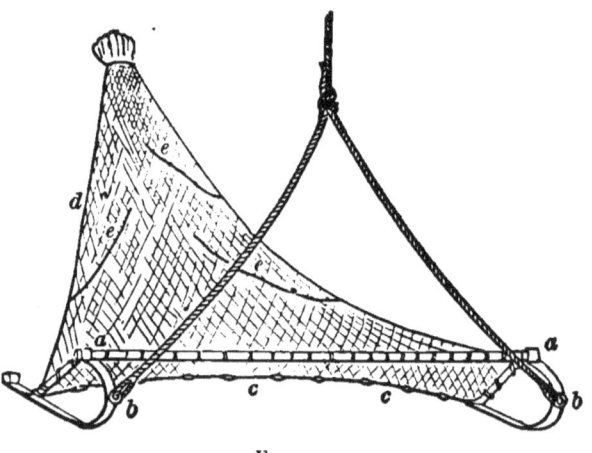

FIG. 14.

marine collecting and is useful for obtaining such bottom forms as are covered with spines. It consists of an iron bar, having on each end a wheel. To the bar are attached about half a dozen chains, each chain having every few inches bunches of hemp or untwisted rope. Such animals as the Starfish, Brittle-stars and Shrimps, become entangled in the bunches and are thus brought to the surface.

Another method of obtaining forms from the bottom is by pumping. This was first employed by Meyer and Möbius in their investigations of the fauna of the Bay of Kiel. A pump was so arranged that the lower end of the pipe was within a few inches of the bottom, and the forms brought up by the current were collected by nets and sieves.

Between the bottom and surface other forms may be found; to collect these a sunken net, first used by Bäur, is employed.

The surface of the sea at times is covered with infusoria (*Noctiluca*, etc.), jelly fishes, larval Echinoderms, Worms and Crustacea, Copepoda, Salpæ, Sagittæ, etc. To obtain these the surface net is employed, fig. 16. This consists of a ring of brass wire about a foot and a half in diameter, to which is attached a net of fine gauze. This is towed through the water, being frequently

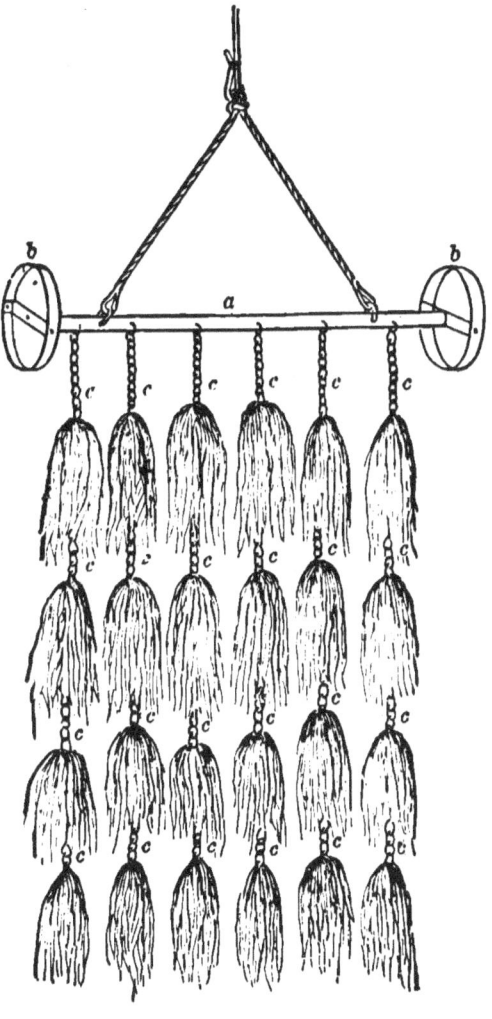

FIG. 15.

pulled in and washed in a bucket of water. On placing this water in the light it will be seen to be filled with microscopic forms. The best time and place for using the net are in protected harbors when the surface is smooth and the sea phosphorescent. A place where two currents meet is especially productive. Surface skimming was first employed by Johannes Müller.

After storms, it is well to examine the beaches to obtain the deep water forms which have been cast on shore. Among the "roots" of the "Devils' aprons" (*Laminaria*)

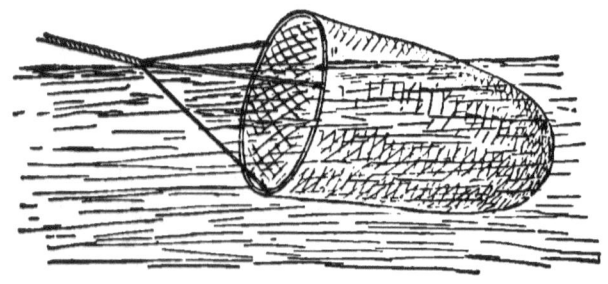

FIG. 10

will be found shells and starfish, while on the fronds frequently occur Sertularians and Bryozoa.

Fish stomachs are another source of obtaining deep-water forms, and they frequently contain rare shells. The fish themselves should be carefully examined for external and internal parasites.

The collections should be cared for as soon as possible, as many forms soon badly decompose. With each package of specimens should be placed a label, written with soft pencil on stout paper, giving exact data of locality, depth, character of bottom, date, etc. The smaller specimens should be placed in homœopathic vials and not mixed with larger forms. For larger forms the common glass fruit jars are convenient, both for collecting and as storage jars.

For the majority of marine forms, alcohol is the best preservative. The specimen should be first placed in weak spirit and after a few hours transferred to stronger and this process again repeated. By this the water is gradually extracted and all undue contraction of tissues avoided. Crustacea and mollusks intended for dissection should have the shell cut or cracked before placing in alcohol so that the spirit may readily penetrate the soft parts which otherwise would rapidly decay.

Medusae and some other forms are not readily preserved without great distortion, owing to the extremely large percentage of water in their composition. Various processes and preparations have been employed, but success is as much the result of accident as of any especial skill or of any superior merit in the preservative. Possibly the best method is to place the jelly-fish for a short time in a one-tenth to one-twentieth per cent solution of osmic acid, and then transfer to fifty per cent alcohol and after a few days place in sixty per cent spirit. Another method is to employ a solution of bay salt of a specific gravity of 1.148, to each quart of which two ounces of alum have been added. The specimen is daily changed to a fresh portion of the solution for a week. Methylated spirit, 30 under proof, with forty drops of creosote to the quart has also been recommended.

It is frequently desirable to preserve animals in their expanded condition. So far as the writer is aware no method has been found in which gasteropods can be so preserved, though many experiments have been tried. Sea anemones, etc., may be readily killed expanded by gradually adding picric acid to the vessel in which they are contained. Another way is to allow them to die in sea water which has become stale. The former method, however, is the most successful. Fresh-water Polyzoa, it is said, may be killed in an expanded condition by adding a few drops of alcohol or brandy to the water in which they are living.

Should it be desired to preserve the shell of a mollusk without the animal, the whole may be macerated in water and the contents carefully washed away. Bivalves should have the two halves carefully tied together, while care should be taken to preserve the *operculum* of such gasteropods as possess it, as it has considerable systematic value.

The same methods may be employed in collecting fresh-water invertebrata as in marine. Ponds and lakes can readily be dredged and a trawl or siene will frequently bring up numbers of rare forms. The beds of rivers contain numerous shells (Strepomatidæ, Viviparidæ, Limnæidæ, Unionidæ and Cycladidæ), for which careful search should be made. A dipper, with a perforated bottom, on a long stick, is frequently a handy substitute for a dredge, in shallow water.

Land shells are most numerous in a limestone country. A good place to hunt for them is under boards or fallen leaves.

LABELLING AND MOUNTING.

CHAPTER II.

LABELLING AND MOUNTING SPECIMENS.

It is on these two points, labelling and mounting, that much of the instructiveness of a museum or collection depends. The labelling conveys the information regarding the specimen, while the mounting places the specimen in the best position for observation and study.

LABELLING.

The labels used should, in size, be in proportion not only to the size and prominence of the specimen, but also in relation to the amount of information to be conveyed. It is best to have but few sizes and to have a certain amount of regularity in the labels employed. The most useful size is one inch by two and one-half inches, but larger and smaller ones must be occasionally used and the sizes of these must be selected by those in charge of the collections. The smallest, except those for insects, should measure not less than one-half by one and one-half inches. For insects a label of one-half by three-fourths of an inch is very useful. The labels of whatever size employed should be as plain as possible and the printing should be confined to a simple border. This border possibly looks best when printed in red ink, and that color is employed for the purpose by most museums. Heavy paper or cardboard is best for the labels. When the label is to be pasted, paper is preferable, but in all other cases the cardboard possesses the greater advantages.

As mentioned above, the purpose of the label is to convey information and this should be expressed in as concise and plain a manner as possible. In some museums (*e. g.*, that of the Boston Society of Natural History), all labels are the product of the printer's art and several copies of each are struck off at once, thus affording a supply from which to replenish as those on the specimens become defaced or injured. The expense for this is far less than would be supposed.

In the majority of cases, however, this plan, cheap as it has been found to be, is beyond the means of museums and hence the labels should be written. This writing should be done with *black* ink and in a legible hand, the ordinary "marking hand" being well adapted for this purpose. For ink, there is nothing better than India ink ground up in acetic acid. Windsor and Newton's liquid India ink is thus prepared and is handiest for the purpose. When it becomes thick by evaporation, it can be diluted by the use of *acetic acid*. Do not use water to dilute it as then the ink is spoiled.

The principal points which are usually to be enumerated on a label are the generic and specific names, locality, date, collector and donor. The adjoined label copied from one in the Boston Society's museum shows the usual form. The generic name should always begin with a capital, but opinions differ as to the initial of the specific, but with Americans the weight of authority seems to be in favor of *always* beginning the specific name with a small letter whether derived from a proper name or not. It should not under any view begin with a capital unless derived from a proper name. Following the scientific name comes the "authority." Here again opinions differ, some claiming that the name of the person who first described the species should be given, no matter whether it belong to the genus under which it originally was described or not. Others claim that the name to be used is

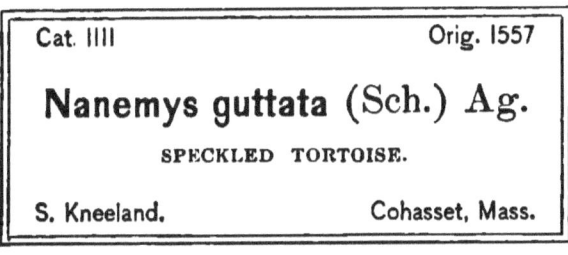

Fig. 17.

that of the person who first used the generic and specific name together. A third party adopt a compromise and give both names together, that of the describer of the species in parenthesis, followed by that of the one who first used the generic and specific combination which is adopted. For instance, Herbst, in 1796, described a hermit crab under the name *Cancer sclopetarius*. In 1852, Dana characterized the genus *Clibanarius*, and in 1859, Stimpson ascertained that Herbst's species should be assigned to Dana's genus. Now according to the first method the name would be written *Clibanarius sclopetarius* Herbst; according to the second

Clibanarius sclopetarius Stimpson; while the latter would be *Clibanarius sclopetarius* (Herbst) Stimpson. The second and third methods are most commonly adopted, the third expressing more than the others. The best authorities omit any comma between the scientific name and the authority.

Should the specimen be a *type*, an abnormal form or immature stage, or possess any important features, that fact should be noticed on the label. The original labels coming with a specimen should be scrupulously preserved in connection with it, as they give a value and authenticity which the specimen could not otherwise have.

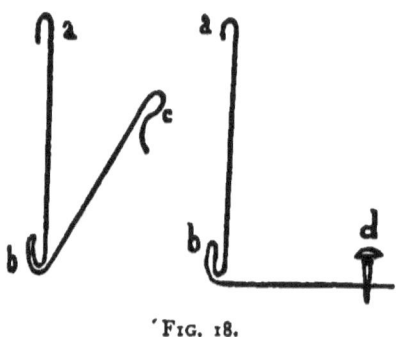

FIG. 18.

Various methods have been devised for affixing labels to specimens. When the object is fastened to a tablet, it is best to affix the label with paste or mucilage, or with short pins, one at each end of the label. A mucilage made of equal parts of gum tragacanth and gum arabic in water, to which a few drops of glycerine and carbolic acid have been added, is possibly as good as any for fastening paper to wood, glass, stone or metal. In all other cases the label holder devised by the late Caleb Cooke is very advantageous. It is readily made by folding a strip of thin tin one-eighth of an inch in width in the manner shown in fig. 18. The label is placed in the folds shown above and below (*a* and *b*) which are then closed with a pair of pincers. The label and holder are then fastened to the object holding the specimen. In the case of a bird stand or wooden tablet, this is effected by perforating the end *c* and then using a tack. When applied to a bottle, copper wire is used. The end *c* is folded around the wire and the wire then placed around the neck of the bottle and the ends twisted tightly. The advantages of this method are many; the label is firmly held and at the same time can be readily removed by loosening the folds with a knife. All original labels can be safely preserved out of sight by placing between the public label and the tin strip. The labels are also kept flat and by bending the tin at *b* can be readily disposed so as to be more easily read when on a high or low shelf.

In the case of alcoholic specimens a label should always be kept in the bottle, as well as one on the outside. This label should be written with a *soft lead pencil*, or with India ink dissolved in acetic acid (never with common ink) on parchment or very strong paper. This label, which is intended only for the student and curator, should contain every item of information regarding the specimens.

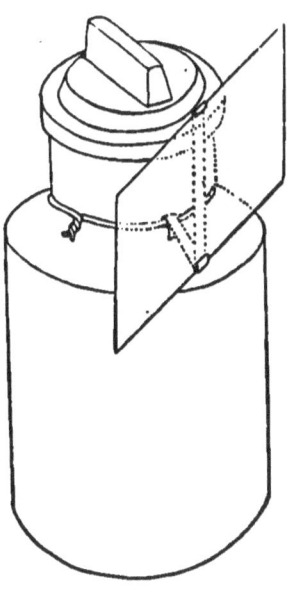

FIG. 19.

Insect-labels should be kept on the pin and should be small. It is usual in arranging insects to place them in rows and then in the farther left hand corner is placed the family label, next comes the generic, and then the specific; the insects belonging to the species are then extended in a transverse row and following them on the left of the case, comes the next specific label and so on. Some have adopted a method of showing the geographical distribution of forms at a glance by the color of the label, and in geological collections the "age" by the same method, but it is doubtful if it repays the extra trouble involved.

Labels can be obtained of "regular" sizes of most dealers in naturalists' supplies at a cheaper rate than they can be produced by local printers. Generic and specific names for labelling certain groups of insects are also kept on sale and well repay the cost.

Catalogues are important, though some of the largest museums almost wholly dispense with them. The best results follow the use of the double system employing both books and cards. In a book prepared for the purpose, each specimen is entered as received, with all possible information. For this the books should be ruled in columns for the following entries: original number, current number, number of specimens, name, sex and age, where collected, when collected, by whom collected, donor, remarks. A number is then affixed to the specimens corresponding to the entry in the book catalogue. Sheets of printed numbers for this purpose are kept for sale by dealers in naturalists' supplies. Where possible, parchment numbers should be used and tied to the specimen. Even better than parchment is the method, which is now extensively adopted, of stamping the number on sheet zinc with the steel punches which may be bought at any hardware store and then affixing the zinc to the specimen with stout twine or small copper wire. It is, however, difficult by any ordinary method to affix a label permanently to a fossil or mineral; strings and wires will become loose and paste and gum will crack off. In such case, the writer has adopted the method of putting on each specimen (in the least conspicuous place) a small spot of white paint, and on this, when dry, the number is written with a pen; there is no danger of such a label being detached and lost. In the book catalogue the specimens are arranged simply according to number and without regard to systematic relations, which are to be found in the card catalogue.

This card catalogue is made of cards arranged alphabetically or otherwise as may be desired, each bearing at the top the generic and specific name and below the desired information. These cards should be about three by five inches; their appearance, etc., are best seen from the following diagram.

ARIUS EQUESTRIS.—BAIRD AND GIRARD.						
Cat. No.	No. Spec.	Age.	Locality.	When col'ected.	Nature.	Collected by
836	1	Adult	Indianola, Tex.	1874	Skull	John H. Clark.
1142	4	Young	Brownsville, Tex.	April 25, 1858.	Alcoholic	Capt. Van Vliet.

By this system of book and card catalogues, it can at once be seen exactly what specimens the museum contains, and also, if the number be preserved, any lost labels can be duplicated. It is convenient to have the catalogues subdivided into groups corresponding to the larger divisions of the animal kingdom, others for minerals, fossils, etc. The cards can then be kept in drawers or trays and any necessary interpolations can be made as desired. Cards suitable for this can be obtained of standard sizes at the Readers' and Writers' Economy stores in Boston, New York and other large cities.

MOUNTING SPECIMENS FOR EXHIBITION.

Mammals and birds designed for exhibition are usually stuffed and mounted on stands. It does not fall within the scope of this work to describe the methods employed by the taxidermist in stuffing skins. It is well enough, however, to reiterate the advice given on another page that the stand employed should be as simple as possible and all mica dust, moss and artificial leaves be discarded as they detract greatly from the appearance of a collection when viewed from a scientific standpoint. On the underside of each stand, all information regarding the specimen should be written with a soft lead pencil, paint or India ink. Skeletons and skulls should be supported on wires firmly fixed at their lower end in a board. Ward's preparations are models in this respect.

Birds' eggs may be kept in the nest in which they belong. Should the nest be wanting, the eggs present a very handsome appearance when placed in paper trays lined with pink cotton wool. The most common method, however, is to mount on wooden tablets. These wooden tablets, which are very generally adopted in museums for specimens of all kinds, should be made of whitewood. The grain of pine shows too plainly while basswood warps badly. It is best to have them made with a depression in which to mount the specimen and an elevated portion on which to affix the label. The form is shown in section in fig. 20. These can be made in long

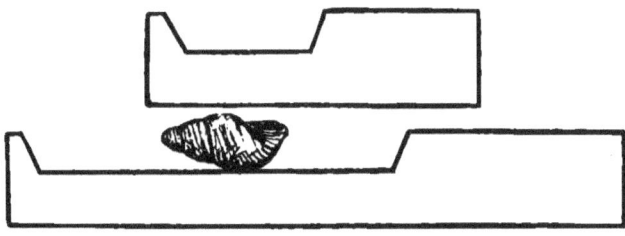

FIG. 20.

strips at any planing mill and then sawed up in lengths to suit. The sizes should be so adjusted that two of the smallest will equal the next in size and so on. They then readily fit together and fill up a case without any bad gaps. The smallest should be one inch wide by two long, the next in size two inches square, then two by four, etc.

These tablets are usually painted dead black. A cheap way of accomplishing this is with lampblack mixed with spirits of turpentine and a very little oil. This, however, is apt to rub off slightly and soil the hands and cases. A more expensive paint, which produces the best results, is the ivory black used by carriage makers. This should be mixed with a little shellac varnish and applied with a brush. Sometimes different colors are used for these tablets; this in the case of geological collections has its advantages, as different formations may be indicated by the color of the tablet bearing the specimen. The general effect, however, is not so good as when a uniform black is employed in all departments of the museum. Various substitutes for the wooden tablets have been proposed and each has its advocates. Pasteboard, slate and glass have all been used, but wood, taking everything into consideration, is the best.

The specimens are affixed to these tablets in various ways. A common method is with "grafting wax." The grafting wax, such as is used by gardeners, is kept melted in a glue pot over a spirit lamp and a drop taken up with a brush and placed in the desired spot on the tablet and the specimen placed on it, and a few moments only are required to set the wax and hold the specimen firmly. If at any time it be desired to remove the object from the tablet, a knife blade slightly warmed will accomplish it, and all wax may be cleaned from the specimen by alcohol or turpentine. Others employ sealing wax dissolved in alcohol, or common glue, for fastening the specimens, but while they hold the object securely, it cannot be as readily removed.

Many minute forms (shells, etc.) cannot be fastened directly to the tablets and at such times small homœopathic vials are used. The specimens are placed in these and the vials are then stopped with a cork or a bit of cotton wool, and then cemented to the tablet by wax or glue. A label should always be placed in each vial. A great desideratum for this purpose is a vial of quadrangular outline with clear flat sides.

In the case of the large branching corals (gorgonias, madrepores, etc.), a square box is taken and filled with plaster of Paris mixed with water; the coral is then placed in this in an upright position and supported until the plaster sets. The box is then taken apart and all rough places smoothed with a case knife, and then the plaster is painted with lampblack and turpentine. This forms a heavy base which holds the specimen uprightly. The label may be applied to this base. The coarse "builders'" plaster should be used for this purpose, as it is much stronger (as well as much cheaper) than the fine.

BOTTLES AND VIALS.

In every museum, vessels of glass are largely employed and form a considerable item of expense. For exhibition purposes, pains should be taken to obtain clear glass, free from bubbles and irregularities producing distortion in the view of the specimen. Glass stoppered vials and jars also are preferable as they not only add to the appearance of the collection, but they are less liable to leak, and permit the alcohol to escape by evaporation than are those with corks. The mouth of jars, vials, etc., should be as large as possible so that specimens proportionate to the size of the bottle may be readily admitted; otherwise a larger amount of alcohol is required than is necessary for the preservation of the specimen.

It is best in the case of minute specimens to place them in homœopathic vials with alcohol, then stop the vial with cork and place the whole, cork downward, in a larger bottle which in turn is to be filled with alcohol. This renders it easy at any time to find the specimen which would not be the case were it loose in a large bottle, while the alcohol in the outer vial will have to evaporate until the cork of the smaller is reached before there is the slightest danger of the contents of the inner bottle being injured.

The best homœopathic vials for museum purposes are those made with straight sides without any neck or shoulder, as then the inside can be readily cleansed and all specimens can be readily taken out for examination. Rubber stoppers do not answer overwell for museum purposes, as the alcohol is apt to affect them and to set free the earth with which they are adulterated, and cover the objects with a dense white precipitate.

In the Museum of Comparative Zoölogy at Cambridge, oval glass jars with flat sides are used for starfishes and ophiurians. The mouth of the jar is ground and covered with a glass plate fastened by cement and also by a strip of tinfoil extending on both the glass cover and the sides of the jar. The specimen is spread on glass or mica plates and fastened with thread, bristles or silvered wire, and the whole placed in the spirit.

Dissections of animal forms are preserved in alcohol by extending on some substance not affected by the spirit. The principal ones employed are mica, glass and wax. The object is fastened to the glass or mica by strings passed through holes bored for the purpose. These holes can readily be bored in glass, with a three-cornered file moistened with spirits of turpentine and mounted in a drillstock. When wax is used the specimen is fixed with insect pins. It is well to blacken the wax by melting it and stirring in lampblack. This forms a good background against which all details are readily seen. Great care should be exercised in selecting the wax, which should be pure. The common adulterations of wax are water, tallow and lard, and the presence of either of these produces a flocculent precipitate in alcohol, which settles on the specimen and ruins it, as it is very difficult to remove.

For storage purposes it is not necessary to use so good a quality of glass as for exhibition. A very useful article is the ordinary fruit jar with glass cover and screw top. The rubber of these jars will occasionally have to be renewed as the alcohol hardens the rubber and renders it brittle. At other times large copper cans are used, fitted with wide openings secured by screw covers, while for the largest forms special tanks of copper or zinc are made. A barrel can be readily fitted up for containing specimens, by carefully smoothing off one end, removing the head and adjusting a wooden cover with rubber packing over the end. To the sides of the barrel are attached iron bars terminated by screws and these project through the lid and by means of nuts fasten it tightly. In any of these large storage vessels, numerous small specimens may be kept by wrapping each (with its label) in millinet, mosquito bar or coarse cotton cloth. The same course may be pursued when sending specimens from the field or from one museum to another. With fishes so sent it is usual to place the labels under the gill covers.

Smaller specimens may be stored in cork-stoppered bottles. A cheap way of obtaining these is to buy the empty morphine and quinine bottles of the apothecary. These are of good glass and have wide mouths. Corks for these may readily be rendered tight by immersing in melted paraffine, or better in paraffine dissolved in benzine. These storage bottles should be so arranged that any desired specimen can readily be found.

One thing that should be constantly kept in mind in the museum is that it is as easy to have too much on exhibition as too little. The primary object of a collection is to instruct, but with many confusion only results. Every specimen should not be on exhibition; nor should every species or genus. It should be the object of the curator to make the collection typical; to select those species which best illustrate the larger groups, while all others are relegated to drawers, boxes, etc., where they will be readily accessible to the special student but will not aid in confusing the average museum visitor.

The space thus gained should be utilized by labels and cards, conveying in plain language the characters of the various groups. It is also well to place in the cases drawings illustrating the structure and growth of the various classes of the animal kingdom. These may be plain or colored according to nature, or conventionally, to show more clearly the details of structure. The following conventional colors are generally adopted :

> Red, heart and arterial circulation.
> Blue, veinous circulation.
> Green, liver.
> Purple, kidneys.
> Yellow, ovaries.
> Orange, testes.
> Brown, alimentary canal.
> Neutral tint, nerves.

Large labels can be easily written with the "Audiographic Pen," invented by Mr. D. S. Holman, of Philadelphia. This has since been extensively sold under the name of "Automatic shading Pen," and can be obtained at any stationery store.

ROOMS AND CASES.

CHAPTER III.

ROOMS AND CASES.

IN a work intended for all classes of naturalists, no definite rules can be laid down to govern in each case the construction of the home of the collections. In many instances, the museum is a private one and is kept in a room of the dwelling house; between this and such immense collections as those of the British Museum and the Jardin des Plantes, every gradation may be found, each requiring peculiar accommodations.

In the case of private collections, a room should be selected, if possible having a northern exposure, well lighted and fitted up with conveniences suitable to the nature of the specimens. The windows should be screened with curtains of yellow "holland" as this color tends to exclude the actinic rays of light and to preserve the specimens from fading.

For larger collections, such as are possessed by most colleges and many societies, more extensive accommodations are necessary and a building should be especially devoted to them. The average college museum building is but poorly adapted for its purpose; it is the result of consulting architects who know nothing of the requisites of such edifices.

The architect draws some showy or striking "elevation" with useless towers and spires and narrow windows, and leaves internal arrangements to chance. The result is an ill contrived building, with inaccessible and useless rooms, numerous dark corners and disagreeable cross lights. But the greatest disadvantage lies in the impossibility of maintaining anything like a systematic arrangement of the collections. The proper way is first to arrange the rooms and apartments and then to accommodate the walls and the roof to them. It would be well for all having charge of the erection of buildings for the display of specimens of Natural History to visit some of the larger museum buildings such as those at Boston, Cambridge, New Haven and Washington and consult with the authorities there in charge as to the advantages and disadvantages of the building occupied by them. It might also be an advantage to visit the museums of New York, Princeton and above all Philadelphia,[1] to see how a museum building should *not* be constructed.

The following plan is here inserted as a hint which might be useful in the construction of a building of moderate size. It contains some features of value but can of course be modified to suit circumstances. It is primarily designed for the use of the average college.

---

[1] The cases at Princeton are (or, were, at the writer's visit) worse if possible than the building, while no museum building could be less adapted for its purpose than that of the Academy of Natural Sciences of Philadelphia. Those collections of Europe which are tucked away in the corners of some old castle or which are displayed in the cloisters of some former monastery are fully as well provided for. The building is the result of architects working without intelligent supervision and was constructed by the Academy in direct opposition to the views of its best scientific members.

It consists of a main portion and a wing each two stories in height. The main portion is a square of say fifty feet. The walls are solid, there being no windows in the sides.

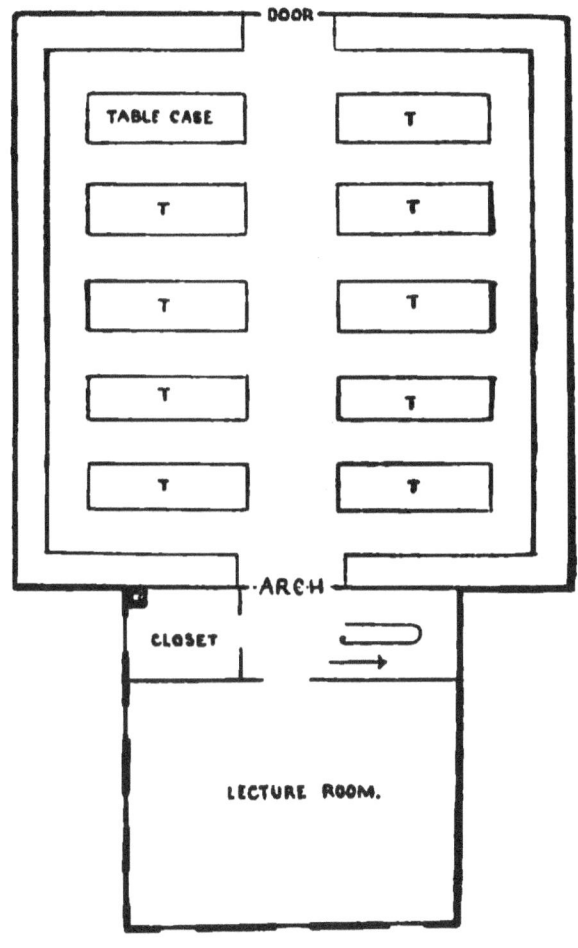

FIG. 21.

The second floor consists of merely a gallery and thus the whole of the main building is a single room, and lighted by a lantern window in the roof. The walls are occupied on both floors with vertical cases, and the lower floor is taken up with table cases which are best for all specimens except mounted vertebrates and alcoholic collections. The gallery is

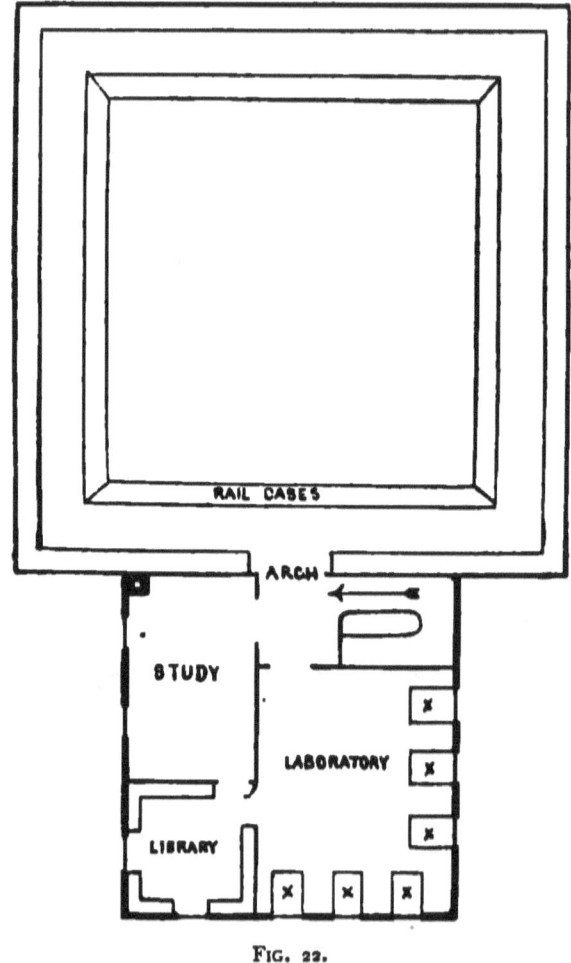

FIG. 22.

surrounded with a railing and this may also bear horizontal cases.

Connected with the main portion by arches, on each floor

is a wing of say thirty feet square. This wing is lighted by windows in the ordinary way and contains on the ground floor a hall with stairs leading to the basement and the second story; a closet for the janitor and a lecture room. On the second floor are the laboratory, a room for necessary books of reference, and a small study for the professor or curator in charge. In the basement can be placed the means of heating; room for the storage of specimens, macerating rooms, etc. If possible, both basement and laboratory should be supplied with water.

For museums of the larger class, this plan will of course prove inadequate and some other must be adopted. The architect and others having the erection of a museum in charge should visit the larger museums and consult with those having them in charge.

A museum building should always, if possible, be isolated and built in the most thoroughly fire-proof manner. The walls should be of brick or stone, the girders, joists, etc., of iron, and the floors of brick, iron, slate or some other incombustible material. Museums are far too valuable to be entrusted to wooden buildings and even those thought to be fire-proof have not always proved so. The fires at Portland and Chicago each destroyed valuable collections stored in buildings which were supposed to be secure against the devouring element.

One feature which should be adopted in every museum building is adequate provision for laboratory work. In college museums this is best accomplished by having a large room where all can be at once under the supervision of the instructor. In buildings for society purposes it is better to have a number of small rooms for this, which may be occupied by those of the members who care to do any work in the building. In the Philadelphia Academy building these rooms are in alcoves leading from the Library; in the Boston Society's Building there are two on each floor leading from the exhibition halls; in the New York Museum they are all on the upper floor of the building.

CASES.

The cases are by no means an unimportant portion of a museum and great care should be taken in their construction. It will not do to leave them entirely to a builder or cabinet maker; a naturalist should also be consulted. From an omission in this respect the cases in many museums are poorly constructed. Notable examples are to be found in the instances of Brown University and Princeton and Williams Colleges. At Brown the cases are very loosely constructed, leaving large holes for the entrance of dust and vermin; at Princeton the extent of sash nearly equals that of glass, rendering it almost impossible to see the specimens on account of darkness; while those at Williams cannot be tightly closed and the shelves are permanent and cannot be altered in height. On the other hand, the cases of the American Museum in New York, Yale College and the Peabody Museum of Archæology at Cambridge are models, but are very expensive. The cases of the Peabody Academy of Science at Salem, Mass., are very good and others can be built like them at a cost not exceeding that of a poor case. It would be well for those having charge of the equipment of a museum to visit these four museums before building their own, and thus avoid the endless grumbling and dissatisfaction which might otherwise follow.

Cases for exhibition may be divided into two groups, vertical and horizontal. The former are generally either placed against the wall or are used to divide the exhibition room into alcoves. Each upright case should have its own floor, the floor of the room never being employed for that purpose. Cases placed against the wall should also have their own back. Otherwise any settling of the building will produce cracks through which dust and vermin find easy access. Both floor and back should be made of thoroughly seasoned matched lumber, or better still of zinc and should be fastened to the rest of the case without the slightest crack. The rest of the case should be of well seasoned timber, as light as is consistent with strength, while large panes of *good* glass permit a clear view of the specimens exhibited. The doors should be very firmly made so that they cannot sag and plenty of hinges should be used in hanging them. Means should be employed of fastening them tightly in at least three places. The lock invented by Prof. E. S. Morse, improved and manufactured by Mr. Jenks at Middleboro, Mass., is admirably adapted for this purpose. In this lock all bolts act as wedges drawing the door closely against the projecting portion of the jamb.

To render the joint between the door and the frame tight, several methods have been employed. Sometimes a thin strip of cotton wool has been tacked to the door, at others the door and frame are fitted with tongue and groove. This is possibly the best method and is employed in the Yale cases. A cheap means has been adopted at the Peabody Academy of Science with good results. A thin strip of rubber packing is folded and fastened by means of a strip of wood to the case and against this fold the door closes sending the joint all but airtight. The construction is readily seen from figure 23. The floor of a vertical case should be some few inches above the floor of the exhibition room, and the space thus left may be occupied by drawers for the storage of specimens. The shelves in an upright case should be adjustable to any desired height. A perfect method of accomplishing this yet remains to be invented. Sometimes sticks fitting into ratchets on which the shelves are supported are employed, others support the shelves on "screw eyes" screwed into the frame of the case. But doubtless the best apparatus is the adjustable brackets. Two patterns of these are made, one by Mr. Jenks and one by Mr. Gavitt, each having its merits and objections. Possibly the former is preferable. Each of these employs an iron bracket (horizontal or inclined as may be desired) which hooks into an

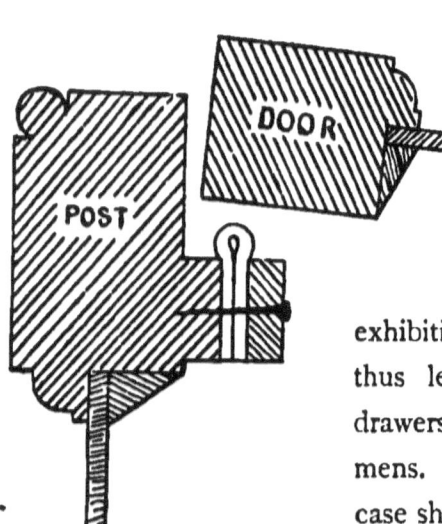

FIG. 23.

iron frame and can be raised or lowered as circumstances demand, by simply unhooking and placing in another hole. These bracket irons should never be affixed to the wall of the case but to a post inserted especially for the purpose.

Upright cases are necessary for alcoholic specimens and mounted mammals and birds; all other forms, with a very few exceptions are better displayed in horizontal or table cases. In the construction of these the same care to make the cases tight should be used as in vertical cases, and the same methods, with slight modification, may be used. The manner of applying the rubber strip is shown in fig. 24. It is best to make the case deep enough to accommodate any specimen that may be obtained, and then to have a false bottom which by blocking up will bring the contents as near the glass as is desired.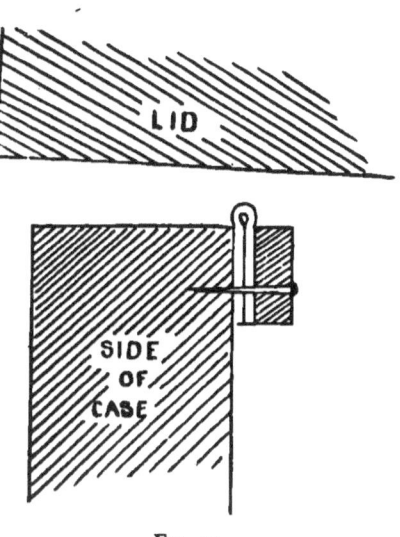

FIG. 24.

It is sometimes convenient to have upright cases in connection with the horizontal and at such times the form shown in fig. 25 possesses advantages. Horizontal cases may be clear underneath, or they may be fitted with glass for the exhibition of large specimens, or with drawers. Specimens placed in such position cannot be seen to advantage; while drawers afford a large amount of storage room and hence in most instances are preferable.

Insects are placed on exhibition in trays which are placed in horizontal cases. These trays are made of light wood and should be about ten by twelve inches by two deep. The bottom should be lined with sheet cork, which may be procured of any dealer in Naturalists' supplies, and over this should be stretched paper so that the whole will present a neat appearance. Prof. E. S. Morse has described in the

FIG. 25.

pages of the American Naturalist a convenient substitute for cork in the bottoms of cases. A rectangular frame of light wood strips of such a size as to be readily admitted into the tray has stretched upon it sheets of paper, one above and one below. The paper may be readily stretched by thoroughly wetting it and while wet gluing it to the frame. Thus, when dry it is as tight as a drum-head. This papered frame should be supported about an eighth of an inch above the bottom of the case. The pin bearing the insect is passed through both sheets of paper until its point penetrates the wood. Thus three points of support are obtained. Other substitutes for cork have been proposed, corn stalks elder pith *pita*, etc., but their use is attended with this disadvantage, that the sap which they contain produces by fermentation an acid which corrodes a pin.

Storage cases for insects are usually made in pairs, and

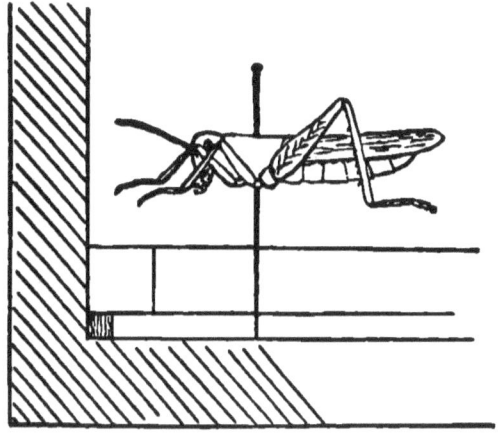

FIG. 26.

should have the two parts fitted together by tongue and groove. This will be very effectual in excluding vermin.

A large proportion of the alcoholic specimens of any collection will be kept in what are known as homœopathic vials and various ways for arranging and keeping these have been devised. The common method is by laying them down in shallow drawers, but this has the disadvantage of injuring the cork by keeping it constantly soaked with alcohol. Mr. Emerton has suggested a handy form which is well adapted for laboratory purposes. Its construction is readily seen from fig. 27. It consists of a shallow tray with a series of steps, the bottles being held in place by rods running along the case. When a row is not full, the vials are fastened by a wedge.

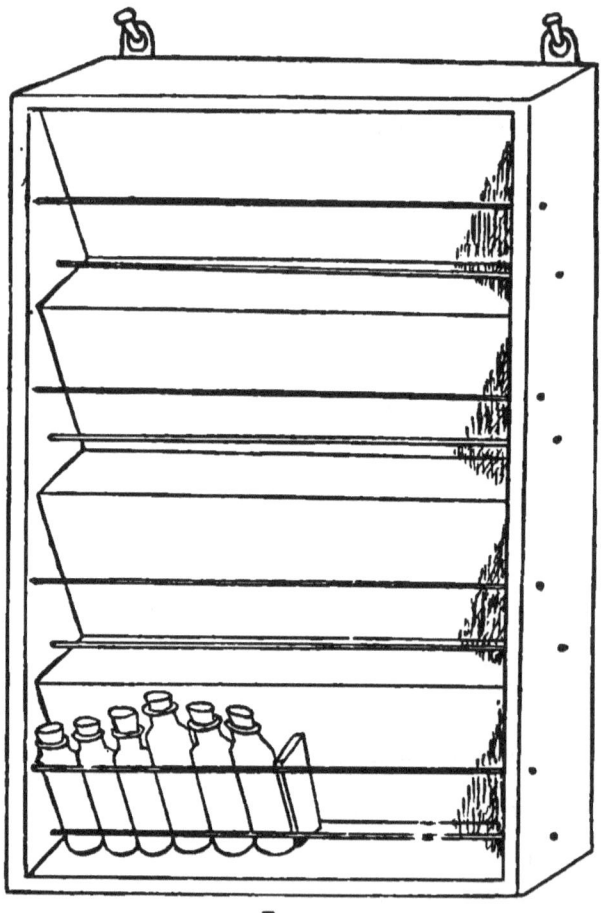

FIG. 27.

This form may stand on the laboratory table or can hang against the wall as desired. They can also be hinged together in pairs, labelled and placed on shelves.

Perhaps the best form of case for homœopathic vials is that described by the same gentleman in the American Naturalist. Narrow deep drawers are made with the front, bottom, back and one side of wood while the other side consists of two wires. This holds the bottles in an upright position and also admits an easy examination of the contents. These drawers may be m:de of varying width but in no case should the front be less than an inch across. This is none too wide for the smallest vials. By making the drawers wider, larger

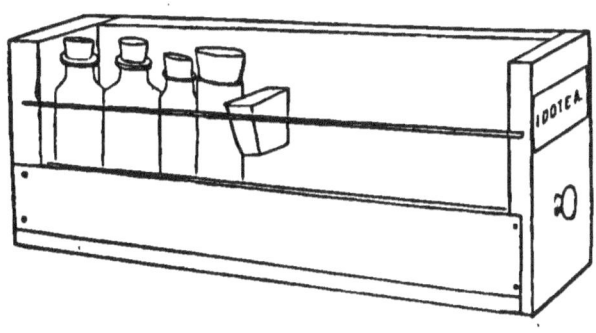

FIG. 28.

vials (one and two oz. can be admitted. The drawers may advantageously be four and one-half inches in depth and twelve inches in length. The bottles are fastened in position with a wedge as shown in fig. 28. These drawers may be placed together in a cabinet (fig. 29) and are interchangeable. By this means any desired arrangement of the collections can be effected, new specimens can be interpolated at any time and by having the drawers labelled any desired specimen can be at once found.

Microscopic slides also require special cases. Of these

there are primarily two forms, those in which the specimens lie flat and those in which they stand on an edge. The for-

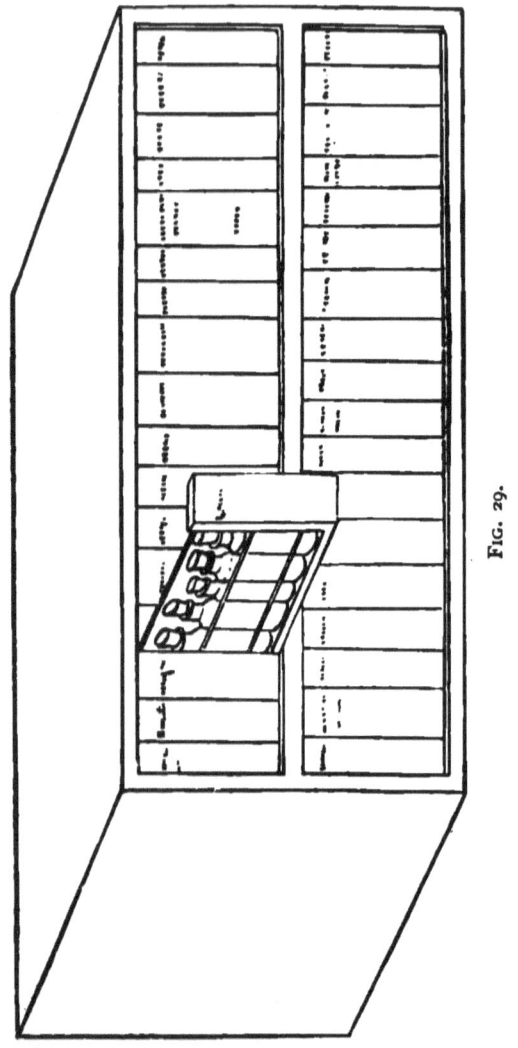

Fig. 29.

mer are preferable from the reason that the "mount" is less liable to slip. On the other hand they take up much more room. The specimens may be kept flat in drawers sliding in a frame or in trays piled one on another and enclosed in a light box; the former is the most convenient, the latter the more compact.

When slides stand on their edges they are supported in the box by strips of wood in which transverse grooves are cut with a saw. In a box of this character many slides may be packed in a small compass. Another method which has been proposed is to take the frame of an ordinary school slate and replace the stone with pasteboard. Rubber cord is then sewed through the pasteboard forming clips which support the slides. Several of the frames are bound together in book form and placed on the shelves.

The forms of cases above described will answer in the majority of instances, but occasionally circumstances will demand something different. No rules can be laid down to cover every condition which may arise; a use of common sense and ingenuity will solve most difficulties.

THE MICROSCOPE.

CHAPTER IV.

THE MICROSCOPE.

To the student of Nature the microscope is indispensable; he requires it to obtain an enlarged view of the objects he studies. The simplest form of microscope is a piece of glass with one or both sides convex, and known as a lens. These

FIG. 30.

simple lenses are very cheap and still very handy. It is sometimes desirable to have two or three so mounted that either one or more may be used, as occasion demands a greater or less amount of amplification. This is frequently obtained by an arrangement similar to that shown in fig. 30. For very low powers it is convenient to have the style of mounting used by watchmakers as this can readily be held by the muscles around the eye, leaving both hands free for work.

With the simple lens there are, however, disadvantages; as, when a clear view is obtained of the centre, objects at the margin of the field are blurred and surrounded by rainbow hues.

Various plans have been adopted to avoid these defects

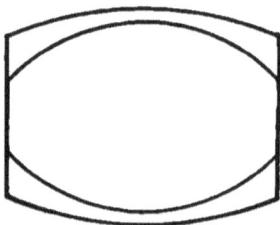

FIG. 31. FIG. 32.

(known as spherical and chromatic aberration), the simplest is that of the Coddington lens (fig. 31) in which the glass has a deep groove cut around the margin which is filled with some black pigment leaving only a small opening in the centre. This to a considerable extent does away with the color and distortion but at the expense of the brilliancy of the view obtained.

Another, and the better, method of avoiding aberration is by having the lens made of different kinds of glass, which produce different effects on light and which tend to balance each other. These lenses are sold under the names of "doublets" (two pieces of glass), "triplets'" (three pieces), "platyscopic lenses," etc. A section of a triplet is shown in fig. 32. A good triplet gives a perfectly flat field and is free from rainbow hues around the object viewed.

Some means of support should be devised to hold the simple lens. A very simple one may be made by means of a block of wood, two bits of stiff iron wire, and a couple of corks. The block of wood should be used as a base. In its centre one of the pieces of wire should be fixed in an upright position. On this wire one of the corks should be

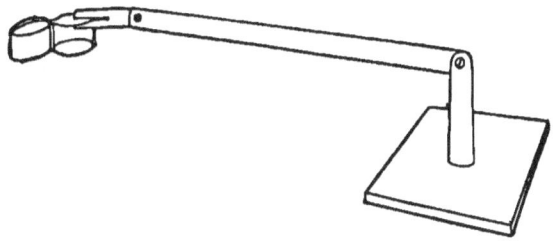

FIG. 33.

made to slide up and down freely, but not so loosely as to slip. Through this cork the second rod should move with the same freedom as the first, but at right angles; and at its farther extremity, the second cork, which is shaped to hold the lens, should be fixed. The corks used in this way afford a very smooth motion and the apparatus produces good results.

A rather more expensive piece of apparatus for this purpose is that shown in fig. 33, and which hardly needs any explanation. The whole is made of brass and is especially adapted for the usual style of mounting triplets. The two joints of the apparatus enable a person to raise and depress the lens and still keep it horizontal. The end portion consists of a piece of brass tube with a slot cut in it to receive the cover of the lens. Such an instrument can be made for two dollars at the most, and to all intents and purposes will serve instead of a dissecting microscope, all forms of which are more or less inconvenient. When it is desirable to use transmitted light, the watch glass or other transparent dish may be placed in the mouth of a bottle and thus sufficient light for all ordinary purposes can be obtained.

Dissecting microscopes are advertised by all dealers in microscopic goods, but they are but little used by professional naturalists, a triplet with a stand answering all their purposes and that with few of the many objections which they all have.

The compound microscope is next to be considered. In this instrument an inverted image is formed by the lens (or combination of lenses) nearest the object (called the object glass) and viewed by the other lens nearer the eye (the eye-piece or ocular). These lenses are mounted in a tube fitted with appliances for bringing them nearer to or removing them farther from an object placed on the stage. Suitable methods are also employed for illuminating the object, and a stand to support the whole completes the list of necessary portions. These will now be taken up in order and their various requisites discussed. In this the writer is well aware that the views advanced are in direct opposition to those held by many *microscopists*, but he is also aware that they are in almost full accord with the opinions of those who use the instrument as a means of research and not as a plaything.

First and foremost, the simpler the working parts are, the better. Complication means a waste of time with no corresponding gain. When a microscope becomes a mass of machinery with screws, wheels, pinions and a thousand and one appliances, its sphere of usefulness is gone.

The tube of the microscope should be short and, if the owner has money for objectives, a draw tube and an amplifier are utterly useless. The use of a draw tube is to increase the length of the tube of the microscope and thus enlarge the image formed by the objective, but it must be at once evident that the increased size of the image is counterbalanced by a corresponding loss in distinctness and brilliancy. But few objectives are made which will stand the strain of the higher oculars and a draw tube. The continental workers adopt the other method of using objectives of greater magnifying power to obtain the desired amplification and it is only necessary to refer to their published figures to show the great superiority of their method. The writer would therefore advise instruments with short tubes, the amplification of the image to depend on the objective.

There should be two methods of regulating the distance between the objective and the object: one by which it can be rapidly increased or diminished, while the other works at a greatly lower rate of speed and thus is suitable for small distances. These are called respectively the coarse and fine adjustments. There have been many plans for regulating the coarse adjustment, but two, however, having any extensive application : by sliding tube, and by rack and pinion. For the novice the latter is the better method but in the hands of the experienced person equally good results are readily obtained by the former plan. With the sliding tube, the tube carrying the lenses is made to slide in a closely fitting collar by a screwing motion. The rack and pinion dispense with the collar and move the tube by a toothed wheel, working into a straight bar fitted with similar teeth. The great objection to this is that the teeth wear rapidly, thus allowing more or less "play" and causing the tube to move by jerks, a serious disadvantage.

The methods employed for securing the fine adjustment are still more numerous. Some move the "nose piece" (*i. e.*, move the objective without altering the position of the tube) ; others move the tube, and still others move the stage. These last forms are the worst of all and should never be employed. Between the other two and the various methods employed for each, there is but little choice when well made. The purchaser should, however, always see that the fine adjustment works easily, responds to the slightest turn of the adjusting screw, is durable, and can be regulated for very short distances. If proper precautions are taken by the maker to secure an absolutely straight motion without any lateral deviation or shake, it is perhaps best to have the whole tube move by the fine adjustment, rather than the objective alone.

The lenses are the important portion of the instrument and upon their perfection its value almost entirely depends. The eye-piece may either consist of two or three lenses mounted in a short tube (Huygenian oculars), or the lenses may be united into one, forming the solid eye-piece. The latter is the better and is at the same time more expensive form. It is best to have two eye-pieces (those lettered A and C by most American manufacturers being the most useful). The greatest attention should be paid to the selection of the objectives and, unless the purchaser be an experienced person, some expert friend should be called in to pass judgment upon the lenses submitted. The great points to be secured are absence of color, flatness of field, and distinctness of image. All objectives above a one-fourth should be provided with an adjustment for cover glass. The "angle of aperture" should be reasonably large, but "high angled" lenses possess no value corresponding to their high price except to "Diatomaniacs."

Objectives are rated by English and American manufacturers according to their focal length, this being the distance between the object and the "optical centre" of the lens when an image is formed. Thus a $\frac{1}{4}$ inch does not have a quarter of an inch between the objective and the object, but between the optical centre (which is nearer the distal portion of the lens than is the mathematical centre) and the object.

For the beginner, the most useful objectives will be an inch and a quarter or fifth. As he proceeds in his studies and his familiarity with the instrument increases, higher, lower, and intermediate powers will be necessary.

Among the higher powers ($\frac{1}{8}$ inch and upward) it is customary to have the lenses of the kind known as "immersion." In these the end of the objective is wetted with a drop of water which forms a thin film between the cover of the slide and the lens. It is thus possible to obtain a more brilliant view of the object as a larger amount of light can be passed through the objective. Besides this there are other immersion lenses, etc., in which oil, etc., take the place of the water.

It may not come amiss to say that the objectives of different makers, of the same nominal focal length, vary greatly in their magnifying power. This results from the fact that some manufacturers, in plain English, *lie* about their lenses and sell for a fourth, for instance, a lens which in reality is a sixth or an eighth, and thus obtain a reputation for making lenses of wonderful power, while were their work tested upon its true merits its rank would be much less. One prominent American manufacturer notoriously does this and upon just this fraud has acquired a great reputation.

Continental manufacturers have adopted an arbitrary system of numbering their objectives, and, for the convenience of many, the tables on the opposite page giving the equivalent of each in inches are inserted.

Other prominent European, as well as the English and American makers designate their objectives by their focal length. The objectives of the Continental manufacturers are fully equal for work to those of English or American opticians while their prices are greatly lower, and the writer would here advise every one to buy the objectives of Hartnack or Zeiss, until American manufacturers offer their work at reasonable prices.

The stage of the microscope should be firm and rigid. It is frequently convenient to have a stage of glass sliding upon brass supports, as thus a great smoothness of motion is ob-

HARTNACK.		ZEISS.		SCHIEK.		NACHET.	
No.	Focal length.	No.	Focal length.	No.	Focal length.	No.	Focal length.
1	2	A	1	1	2	0	2
2	1	A	$\frac{3}{5}$	2	8	1	1
3	$\frac{3}{4}$	B	$\frac{2}{5}$	3	$\frac{1}{2}$	2	$\frac{1}{2}$
4	$\frac{1}{2}$	C	$\frac{1}{4}$	4	$\frac{1}{4}$	3	$\frac{1}{4}$
5	$\frac{1}{4}$	D	$\frac{1}{6}$	5	$\frac{1}{8}$	4	$\frac{1}{5}$
6	$\frac{1}{5}$	E	$\frac{1}{8}$	6	$\frac{1}{6}$	5	$\frac{1}{8}$
7	$\frac{1}{6}$	F	$\frac{1}{15}$	7	$\frac{1}{9}$	6	$\frac{1}{10}$
8	$\frac{1}{9}$			8	$\frac{1}{11}$	7	$\frac{1}{14}$
9	$\frac{1}{11}$			9	$\frac{1}{12}$	8	$\frac{1}{15}$
10[1]	$\frac{1}{16}$			10	$\frac{1}{18}$	9	$\frac{1}{20}$
11	$\frac{1}{18}$			11	$\frac{1}{25}$	10	$\frac{1}{30}$
12	$\frac{1}{21}$					11	$\frac{1}{40}$
13	$\frac{1}{25}$					12	$\frac{1}{50}$
14	$\frac{1}{28}$						
15	$\frac{1}{30}$						
16	$\frac{1}{40}$						
17	$\frac{1}{45}$						
18	$\frac{1}{60}$						

tained. On the other hand, the glass stage is as frequently in the way, and on the whole the student can very well dispense with it, as its disadvantages will nearly or quite counterbalance its convenience. No rubber should be employed around the stage nor in fact anywhere around the microscope.[2] The

---

[1] Nos. 10 to 18 are immersion.

[2] Besides its electrical qualities which render it a nuisance, rubber is readily affected by turpentine and benzole which are so necessary in *microscopic* work.

under side of the stage should be bevelled around the central opening to admit of oblique illumination, and it is often convenient to have a thread cut in the opening itself to admit of using objectives as "condensers" in using high powers. The stage should also be provided with clips to hold the slide in any desired position. Stage forceps are more bother than they are worth.

The illumination of the object is accomplished by a mirror and by a "bulls eye"; the mirror is supported beneath the stage and should be so arranged as to be readily placed at different distances from the object and also so that the light can be thrown at various angles upon the slide. Two mirrors, one plain and the other concave, are usually furnished so that varying intensities of light may be employed. With high powers a lens is frequently employed to add to the illumination and is interposed between the mirror and the stage. This is called a condenser. Some microscopes have the mirror so arranged as to swing above the stage and thus illuminate opaque objects; in others this illumination is effected by the "bulls eye" a large lens of common glass mounted on a separate standard.

It is usual to have some method of cutting off undesired rays of light coming from the mirror. This is accomplished by having apertures of various sizes so arranged that they may be brought beneath the object. Various methods are adopted to accomplish this but it is difficult to say which is best. The microscope as described with its base, its supports, and its means of connection of the various parts form what is known as the "stand," and this will now be considered.

The stand should be solid and firm, without springiness or "give" in its various parts. The base should be heavy so as to prevent its easy overthrow. It is frequently convenient to have the instrument so arranged that it may be inclined; but if an instrument with short tube and low body be procured and fitted with a camera admitting of use in a vertical position, inclination is rarely necessary.

The accessories which are necessary for the biologist are extremely few. First among them comes the camera lucida or other means of seeing the object and the point of the pencil at the same time. The simplest form consists of a bit of thin glass so mounted near the eye-piece of the microscope that the eye can see the point of the pencil through it, and at the same time the image coming through the eye-piece is reflected by it to the eye. In other forms prisms of various shapes replace the thin glass, or a very small metallic mirror is employed. A prism properly mounted forms the most satisfactory camera.

Occasionally, in differentiating certain structures in the living animal, a polariscope is useful. This consists of two prisms of Iceland spar properly prepared and placed, the one below the object and between it and the mirror, the other, either in connection with the objective or the eye-piece, between the object and the eye. When either of these is revolved around the axis of the instrument, many structures are seen to present different colors which vary as do the relative positions of the two prisms.

Other accessories such as mechanical stages, spot lenses,

Lieberkuhns, parabolas, etc., etc., are but rarely used by the true student and need not be described here.

This is a good place to say a word about the "Novelty," "Globe," "Craig," and other microscopes which are extensively advertised and as extensively recommended by clergymen, teachers, and others. These microscopes, furnished for twenty-five cents, are said to magnify 10,000 times, to show animalcules in water and various other wonderful things, but they are merely catch-pennies, and the clergy who recommend such worthless instruments are entering a field in which they are perfect ignoramuses. These cheap microscopes are poorly made, give distorted and misleading images, and in a word are worse than useless. The Craig is perhaps the worst of the lot.

DIFFERENT FORMS OF MICROSCOPES.

From the days of Adams, Baker, Trembley, and the older investigators, microscopes have been used extensively by naturalists, and of course in these years various styles of instruments have originated, but all forms now manufactured may be roughly classified under two heads, the English and the Continental patterns. The latter are almost always small, of great simplicity, and those of the prominent makers like Zeiss, Hartnack, Merz, or Nachet, are invariably of good workmanship. These have either a circular or horseshoe base from which arises the support of the working portions of the instrument. The stage is almost invariably of brass, without glass or mechanical attachments. The tube is usually supported by an arm or bar and the coarse adjustment is effected by means of a sliding tube. The fine adjustment on all foreign instruments which the author has seen has invariably been well made and moves the arm and with it the tube. The English model is larger and much more complicated and clumsy. The base is usually of the tripod form and the uprights supporting the working parts are much taller than is necessary. In the higher priced instruments the stage usually bears a plate of glass which in turn supports the object. This glass stage, *theoretically*, is a great convenience as it affords a very smooth motion and preserves the working parts from corrosive liquids; but in practice it is a great nuisance and can well be dispensed with. The stage in most of the English models is larger than in the continental and in this respect is better. The tube is generally supported by a curved arm and the coarse adjustment effected by rack and pinion. The fine adjustment indifferently moves either the whole tube or just the nose-piece, many manufacturers making both styles. The tube itself is almost always unnecessarily long and this defect is increased by a draw tube. When English and American students learn that definition is better than amplification, and that the shorter an instrument is, the better and more useful it is, then, and not till then, may we hope for a change for the better in this respect. It may seem out of place in a work of this character to speak of one instrument in higher terms than of another, but there are many who wish to purchase microscopes who have not had the necessary experience to select for themselves, hence the following words are written, and must not in any way be considered as an advertisement, except such as the merit of the various instruments themselves demands.

In the writer's opinion, one of the best stands for all ordinary work is the smaller ·compound microscope manufactured by Carl Zeiss of Jena and designated by him as "V a." This stand alone costs ninety marks ($22.50) and when furnished with four eye-pieces and three objectives, A, C, D, F, giving powers of 20–1500 diameters, sells for three hundred and twenty marks ($80.00). This instrument will answer all the requirements of the naturalist or histologist in any special investigation. The ordinary student, however, does not need these higher powers, and the same stand with three eye-pieces and the objectives "A" and "D" (1 inch and ¼) will answer all ordinary requirements and is sold for one hundred and seventy marks (about $42.50). Zeiss's American agent is F. J. Emmerich, 138 Fulton St., New York, who imports, charging 50 per cent. to cover freights and duties.

The instruments of Hartnack are fully equal in value to those of Zeiss and the differences in price are very slight. Geo. A. Smith & Co., 149 A Tremont St., Boston, are the American agents of Hartnack, or rather, of his successor, Prazmowski, and furnish his instruments at very reasonable prices. The stand III a with two eye-pieces, and objectives 4 and 7, giving powers from 50 to 450 diameters, is sold for

$50.00; with the addition of objective No. 9, the price is $70.00. No better instrument for actual work can be bought. When we come to speak of the comparative merits of the instruments manufactured or extensively sold in America, it is a rather more delicate matter to decide between them, though no corresponding difficulty exists. Beyond all doubt the best stand for the student is the American Histological stand manufactured by J. Zentmayer, of Philadelphia, and with an $\frac{8}{10}$ and a $\frac{1}{5}$ objective is sold for $50.00. Those who prefer a rack and pinion can obtain from this maker essentially the same instrument with this addition; the same stand for $58.00. Were the stages of these instruments an inch lower and an inch larger, as they could readily be made, they would be much more convenient.

R. and J. Beck, of Philadelphia, make an excellent instrument, the "Economic"(No. 263), which with two objectives (1 in. and $\frac{1}{4}$ in.) they sell for $40.00. The same with rack and pinion and two eye-pieces(No. 264) is advertised for $55.00.

Bausch and Lomb, of Rochester, make the "Physicians'" microscope, of fair workmanship, which with two eye-pieces and two objectives, $\frac{3}{4}$ and $\frac{1}{5}$ (No. 550), brings $60.00. Their instruments, however, would be much better did they avoid the use of rubber in their construction.

The instruments and objectives of Tolles, of Boston, possess no advantages at all commensurate with the greatly exorbitant prices charged for them.

In case the student desires higher powers than those enumerated with the foregoing instruments, it will be for his advantage to import the lenses of either Hartnack or Zeiss and have them fitted by an "adapter" to his microscope. By this method, he will obtain *good* objectives at about half the prices charged for similar lenses of no better quality made by American opticians.

Binocular microscopes have of late been extensively advertised, but for work possess not the slightest advantage and are only manufactured so that the makers may add to the prices and to the profits made on their instruments.

THE USE AND CARE OF THE MICROSCOPE.

It is a difficult task to give directions for the use of the microscope as the varying uses to which it is put require as varied a method of handling. All objects for the microscope should be mounted either temporarily or permanently on a glass slide. If it be a moist tissue or an object taken from the water, a drop of water should be placed on the slide, the object placed in it and the whole covered with a piece of thin glass. The slide and its object are now ready for examination. The microscope should now be made ready and the objectives screwed on. It is best to use first the lower powers and then the higher if necessary, with either reflected or transmitted light according as the object is opaque or transparent. There are several advantages connected with this method of treatment, one being that in this way a general idea of the structure is first obtained and the various details are studied afterward. It is also much easier to find an object under a low power, and, placing it in the centre of the field, it is in position when the higher powers are employed.

In focussing the microscope it is better first to run the tube down toward the slide to within the focal limits of the objective, watching the operation from the side and seeing that the cover glass is not touched. Then, with the eye to the eye-piece of the microscope, the tube is slowly moved back by the coarse adjustment until a good view of the object is obtained, and then the fine adjustment is used.

In using immersion objectives a drop of water is put on the front of the objective which is then placed on the microscope and run down to the cover glass so that the water forms a thin film between the objective and the cover glass. A very simple experiment, for which I am indebted to Mr. Phin,[3] shows how the immersion aids in the defining power of the microscope. "Take four ordinary plate glass slides and place a very small drop of water in the centre of three of them. Across the ends of these three slides lay a narrow strip of stout writing paper, and then place the four slides together so that between every two there shall be a drop of water and also two slips of paper to keep them apart. If you now look through these four slides at any object, the spots where the three drops of water have been placed will look like a hole it will appear so clean and transparent." The water produces exactly the same effect with the immer-

---

[3] Practical Hints on the Selection and Use of the Microscope, p. 38, by John Phin, N. Y., 1875.

sion objective. Most objectives use water for an immersing medium but some are adapted for oil of cedar, glycerine or other liquid. Immersion objectives are valuable for some special purposes, but for ordinary work the "dry" lenses are much better.

Most high powers of American objectives have an adjustment for thickness of cover glass. This is only necessary for objectives of very high angle (and the higher the angle, beyond a certain point, the more useless the objective). It is better as well as much cheaper to purchase objectives without this adjustment and then use the thinnest cover glasses made. These objectives without adjustment are always well corrected and give good results.

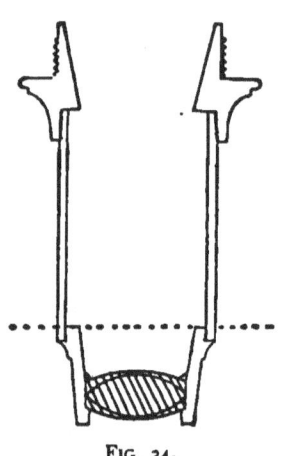

FIG. 34.

Having mentioned high angle lenses above, it may not be amiss to speak of these glasses. This expression refers to the angular aperture of the objective, or, in other words, the angle which the extreme rays of light make with each other in entering the objective, and the larger the angle the "higher" it is.

With every increase in the angular aperture, a shorter working distance of the lens is necessary, and so with very high-angled lenses the inconvenience of working far more than counterbalances the gain in definition. Some makers absurdly claim to make objectives of 180° angular aperture ! People making such claims should be carefully avoided, as their statements are apt to be equally unreliable and false in other particulars. As will be seen from the above it is advisable to purchase the lower-angled lenses. The most convenient glass ever used by the writer was a ¼ of only 48° angular aperture.

One criticism which the writer would make on the ordinary objective is the utterly disproportionate length of brass to the optical portion of lenses, making it next to impossible to use low power objectives on the smaller stands. Fig. 34 represents one of these lenses drawn from measurements, the shaded part indicating the optical portion and the dotted line showing to what extent it might conveniently be shortened. We commend it to the attention of opticians.

USE OF THE CAMERA.

The camera lucida, or camera, as it is commonly called, is one of the most useful microscopic accessories. Two

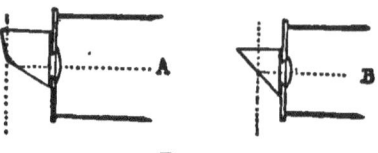

FIG. 35.

forms are offered for sale and are represented in fig. 35; the one at the left, A, is the most valuable one, but B is the cheaper. The ray of light coming through the microscope falls upon an inner surface of a glass prism and from thence is reflected directly to the eye as in B, or to a second surface and thence to the eye as in A. At the same time a ray of light coming from the paper beneath passes through the prism and enters the eye in the same direction as the first. These lines of light are represented by dotted lines in the figures. It will thus be seen that the eye perceives the object under the microscope and a paper placed beneath, superimposed upon one another. Its method of use is as follows: the microscope is first focussed upon an object and then the tube is brought to a horizontal position, the camera attached to the eye-piece and a paper placed on the table beneath the camera. Then looking down through the camera, one perceives at the same time the paper and the object. A pencil may now be made to trace on the paper the outlines of the object, and the details afterward filled in free hand.

In the first figured camera the image by the two reflections is first reversed and then reversed again, bringing it the same as that seen by direct observation through the tube, giving an outline, the details of which are easily filled in.

The second form having but a single reflection produces a reversed image, and hence it is difficult to complete free hand. Instead of a prism, the student may easily make his own camera of this second form by mounting a piece of thin glass in a bit of cork which can be affixed to the tube of the microscope.

A third form, which is rarely seen in America, but which is in the opinion of the writer the best, is the "vertical" camera, manufactured by Carl Zeiss. This form may be applied to the microscope when in a vertical position, throwing the image to one side of the stand, and giving the clearest image of any, of both object and pencil. It is also remarkably cheap costing only twenty-one marks ($5.25). Most students at first find it difficult to use a camera, but practice soon overcomes the difficulty and enables one to adjust the illumination properly, which otherwise is the principal cause of poor success.

The other accessories for the microscope which are of value to the biologist are a stage micrometer, one or more animalcule cages, glass cells, a compressorium (the one described by Mr. Ryder[4] possesses many advantages and is on the whole the best), and a bull's eye condenser which is necessary with opaque objects, and when using the higher powers. The polariscope is of use in mineralogy and in differentiating some animal tissues. Rotary stages, unless accurately centred, are useless and then are of value only to the mineralogists. The stage forceps which accompany most stands are models of clumsiness, are wholly worthless, and can well be dispensed with.

In writing the foregoing pages on the microscope and its accessories, the writer has had this object in view : to show that the simpler an instrument is, other things being equal, the better it is, and that none need be deterred from purchasing a microscope on the grounds that a good instrument costs an outrageous amount. Just as soon as American students realize that the simpler their apparatus is the better their work will be, just so soon will American science rise from its present low level.

There are three men in America, who never did a single stroke of original work in their lives, to whom we are in-

---

[4] American Naturalist, xiv, p. 691. 1880.

debted for the present low state of the microscopical branches of biology. These men without knowing the first requisite of biological work have denounced good instruments and advocated useless ones, have based their opinions of the worth of an instrument on its power of resolving diatoms, and the result is that the beginner, influenced by their dogmatic assertions and widely-copied opinions, has been led to purchase an instrument wholly unsuitable for his ends and the result has been discouragement and a cessation of microscopic work.

THE LABORATORY.

CHAPTER V.

THE LABORATORY.

EVERY museum should have connected with it a laboratory for practical work. This is especially desirable in connection with College Museums, as there is but one way in which Zoology should be taught : directly from the specimens. There are in the United States some 370 institutions which rejoice in the name of college or university, but not ten per cent. afford their students the slightest facilities for practical work. It is sincerely to be hoped that the day is not far distant, when this condition of affairs will be changed and specimens will replace the text-book instruction.

The laboratory should be a commodious, well-lighted room, with, if possible, a northern exposure, and furnished with every convenience for the student. Tables for the students should not be varnished, as in that case any accidental spilling of alcohol will render them sticky and unpleasant. Instead, the tables may be oiled and thus they will not be stained and may be readily washed. Either pine or cherry is a good wood for tables. A convenient size for tables for single students is four feet by three, and two and one-half feet in height. Should it be necessary to place more than one student at a table, the length should be increased so that each may have at least ten square feet of table room.

The support of the table as well as the floor of the laboratory should be firm, so that all unnecessary vibration, which would prove very annoying in microscopic work, may be avoided. In the laboratory of Prof. Alex. Agassiz, at Newport, each table has a support of its own wholly unconnected with the floor of the room.

Concerning the chairs to be used in the laboratory but little can be said. It is best, however, to use either wood or leather-bottomed chairs. Of course no varnish should be used upon them.

Larger tables than those mentioned above should be provided for the dissection of the larger forms, and these should have either slate or metal tops to prevent the fluids, etc., from sinking into the wood and causing disagreeable odors by their decay.

The order should be enforced that every student should put his table in good order at the close of the day, should dispose of all refuse and clean all instruments before leaving the laboratory.

If possible, the laboratory should be provided with water and gas, and there should be kept in close connection a well selected library of morphological works to which the students should have unrestricted access, but should not be allowed to remove from the building. A list of indispensable morphological works is given at the end of the next chapter. Good bibliographies of anatomical and embryological works and papers will be found in Balfour's Embryology, Owen's Anatomy and Gegenbaur's Anatomy.

Besides the tables and chairs, many of the following instruments, apparatus and reagents will be useful, all coming into use in a laboratory where much original investigation is carried on, while in an ordinary college course many may be omitted.

INSTRUMENTS.

Scalpels, large and small.
"Eye knives."
Dissecting scissors, straight and curved.
Forceps of various sizes and shapes.
Microscopes, dissecting and compound.
Hand-lenses.
Microtome.
Beakers.
Tiles, white and black.
Test tube.
Turn tables.
Glass and earthen vessels of various shapes and sizes.
Thin glass.
Glaziers' diamond.
Dropping and dipping tubes.
Tubs.
Aquaria.
Funnels.
Paper.
Pencils

Tenotomes.
Cartilage knives.
Bone saws.
Bone forceps.
Dissecting needles.
Valentine's knife.
Injecting apparatus.
Florence flasks.
Evaporating dishes.
Glass tubing.
Watch crystals.
Hot stage.
Washing bottles.
Water bath.
Glass slides.
Micrometer scale.
Writing diamond.
Thermometers.
Brackets.
Sponges.
Filter paper.
Dissecting forceps.
Colors.

REAGENTS.

Alcohol, absolute, 95 and 50 per cent.
Caustic potash.
Aqua ammonia.

Acid acetic.
" carbolic.
" chromic.
" formic.
" hydrochloric.
" lactic.
" nitric.
" osmic.
" picric.
" sulphuric.
Nitrate of silver.
Extract of logwood.
Alum.
Picrocarminate of ammonia.
Canada balsam.
Dammar varnish.
Ring varnish.
Glycerine jelly.
Marine glue.
Chloroform.
Neutral salt solution.
Asphalt.
Brunswick black.
Bichromate of potash.
Wickerscheimer's solution.
Lampblack.
Caustic soda.
Benzole.
Magenta.

Morphia sulphate.
Curare.
Arsenic.
Corrosive sublimate.
Glycerine.
Carmine.
Eosin.
Vermilion.
Hæmatoxylin.
Creosote.
Prussian blue.
Klcluenberg's hæmatoxylin.
Benzole balsam.
Dammar lac.
Bell's cement.
Gelatine.
Müller's fluid.
Salt.
Chloride of gold.
Paraffine.
Borax.
Acid nitrate of mercury.
Aniline green.
Iodine.
Turpentine.
Indigo.
Oil of Bergamot.
Oil of cloves.

Each student should provide himself with the most useful of instruments and reagents in the foregoing list, while those which are but rarely used might be furnished by the laboratory.

The uses of most of the various instruments, etc., in the foregoing list will be described under the various heads which follow, while the way in which many of the reagents are made will be found in the chapter entitled "Recipes, Formulæ, and Useful Hints."

Each student should make extended notes of all of his work and should accompany it by illustrative drawings. From an experience of several years, the writer regards "note-books" as the poorest form in which to keep notes, as in a short time several books are filled and it becomes an interminable job to find any desired item.

A far better way is to keep the memoranda, drawings, etc., on separate sheets which can be arranged in portfolios and envelopes after any desired system, thus greatly facilitating reference and admitting of future interpolations.

A word in regard to drawing may not come amiss. Most persons have an idea that they cannot draw or learn to draw. Nothing possesses less of truth. Any one with a little practice can make an intelligible drawing, though but few acquire that skill and facility which are necessary for book illustration. Almost every student whom the writer has seen enter a biological laboratory, has said that he or she could not possibly draw and never could learn how. But those same students in a very short space of time would produce creditable drawings to illustrate their dissections. The great secret of drawing is "patience." Drawing takes time, and the trouble with beginners is that they want to hurry. No instruction is necessary to enable a student to reproduce with more or less accuracy the features of any preparation or dissection; practice alone will do it.

Drawings will express far more than pages of description, and whenever it is practicable they should be employed.

For scientific work "bristol board" and a "six H" lead pencil produce the best results. The shades may be put in with India ink and a camel's hair or, better, a sable brush. In case it be desired to color a drawing, water colors are best, and the moist water colors are the most convenient to use.

It is frequently desirable to use certain colors for certain organs and thus through a series of drawings to indicate the parts with similar functions and the following list embraces the conventional colors most used.

> White or neutral tint, nerves.
> Red, heart and arteries.
> Blue, veins.
> Brown, the alimentary canal.
> Green, liver.
> Purple, renal organs.
> Yellow, female sexual organs.
> Orange, male sexual organs.

In a series of drawings with these conventional tints the eye readily appreciates the principal features of the anatomy without the aid of descriptive text. Other organs than those enumerated may be left blank or colored according to the fancy of the artist.

The various photographic processes of reproducing illustrations have lately acquired great prominence and a few hints on preparing drawings for the photographer may prove of use.

The "direct transfer" process of the Heliotype Company is but poorly fitted for scientific work and the results are very unsatisfactory. The drawings are made upon bristol board with an ink containing alum and these are given to the company who produce facsimiles, but they are always muddy and blurred.

For all other photo processes the drawings require that each line should be perfectly black and smooth. The drawings should always be made on bristol board. No wash tints or pencil work will take, but all shades have to be expressed either by lines or dots. Winsor and Newton's liquid India ink produces good results, especially if more cake ink is rubbed up with it.

Of the photo processes the photo-lithographic is the most satisfactory but this cannot be used along with press work, but requires separate plates. The various processes for producing raised plates (photo-electrotypes) do not vary much and the chief distinction between them seems to lie in the skill of the operators. The writer has noticed, however, that by whatever process, if a poor electrotype resulted, it was always attributed to the fault of the one furnishing the drawing, and not to any fault of the photographer, electrotyper or of the process.

LABORATORY WORK.

CHAPTER VI.

LABORATORY WORK.

THE account which follows is from the necessities of the case greatly condensed, many points of great importance being entirely omitted. This account is intended for the beginner only. For more extended directions the student is directed to the list of books at the end of this chapter; those of Huxley and Martin, Tulk and Henfrey, Burden-Sanderson, Stricker and Ranvier giving the best and most detailed instructions. Beale's book, like most of his other works, is in many respects unreliable.

DISSECTING.

All small objects should be dissected under water or a mixture of alcohol and water, as these media tend to support and float the parts and tissues which otherwise would mat together to a greater or less extent, and thus obscure the dissection. Dissecting troughs are used for this purpose. These are usually made of tin (fig. 36). These troughs, for ordinary work, should be about six by eight inches square, and one and a half to two inches in depth. Small tin slips should be soldered to the sides near the bottom to hold the false bottom in position. This false bottom may be of cork, wax, or other material which will hold a pin. It is best for most purposes to have the bottom black, either by mixing lampblack with the melted wax before it is run in, or by painting the cork. Other larger and smaller tanks should be provided for other work. The object to be dissected should be pinned out upon the wax, and just enough water

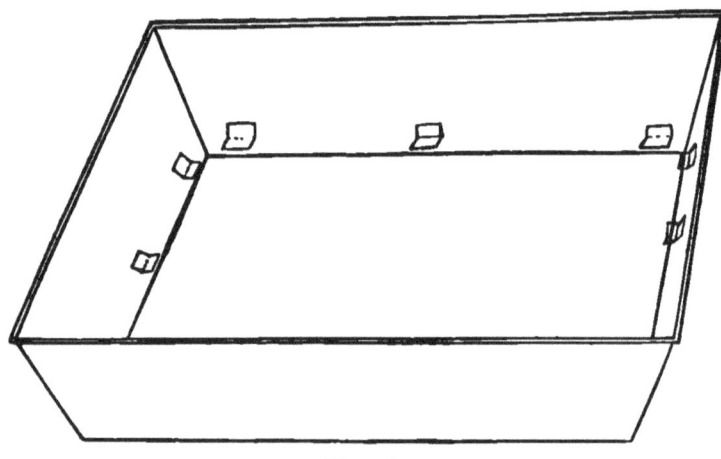

Fig. 36.

to cover the specimen poured in ; a greater quantity interferes with facility of work. When the water becomes discolored or dirty, it should of course be changed.

In case of the smaller animals, as grasshoppers, etc., it is frequently difficult to pin the subject out, but the following plan will answer well. With a hot wire melt a small groove in the wax bottom and, while the wax is still in a molten condition, place the object in it and, when cold, it will be held firmly. Before doing this the specimen should have all moisture removed from the surface, and, of course, it should not be placed above the middle line in the wax.

At the close of the day's work the specimen should be removed from the tank and placed in alcohol, while the tank itself should be placed wrong side up to drain.

Below is given a very brief account of the methods of studying the gross anatomy of a number of types; all descriptions of the organs being intentionally omitted. As the student proceeds, he should make drawings as well as notes of his work and should endeavor to rely as much on himself and as little on text books and instructors as possible.

Protozoa can only be studied with the microscope.

Sponges are best studied by cutting sections and examining them under the microscope. The various structures and arrangements of parts can then be made out more or less clearly, and the eggs and embryos in various stages of development can frequently be seen. To study the spicules, the specimen should be macerated in water, then picked to pieces with needles and examined under the microscope.

Sea Anemones are dissected from the side, when the genitalia, mesenteries and digestive portions are seen. By freezing and cutting transverse sections, the relations of the mesenteries to the alimentary canal are made out. The various cell layers should be studied in stained microscopic sections and the lassoo cells should be looked for.

Starfish should have the upper surface of the arms removed, taking care that the portion around the madreporic body be left uninjured until it becomes necessary to cut it away. *Sea Urchins* may be divided into two halves by a horizontal plane, or one side of the skeleton may be broken in. *Holothurians* should be first carefully examined, and the genital opening, which is near the mouth, found. This genital opening should be placed in the median line above, and then a longitudinal incision made from the genital pore to near the anus. This will expose the viscera and the parts will show a bilateral symmetry. The relation of the longitudinal nerves and canals, and the ambulacra should be studied by cutting through the integument and one of the longitudinal muscles.

Clams are dissected by removing one valve. In order that uniformity may be obtained the valve removed is the left one. To ascertain the right and left, hold the clam with the hinge from you, and the end from which the siphons extend at your right hand; the upper valve will then be the left one. Insert a dull knife in the gape of the shell and cut the strong muscles which hold it closed. These will be found in the clam near the hinge line, at the two ends of the shell. In the mussel (*Mytilus*) and the oyster but one such muscle will be found. The heart (near the hinge line), the alimentary canal with its tortuous course and the nervous system may be then studied. The gills under the microscope show a fine example of ciliary action.

Snails should be extracted from the shell, by breaking it or otherwise, and opened from the dorsal surface.

The larger worms may also be opened from above, but many of the smaller ones, especially among the lower forms, must be studied in sections.

Lobsters are opened by removing a portion of the carapax exposing the circulatory apparatus, etc. The nervous system lies on the floor of the body cavity and in a portion of its course is covered by bony arches. These must be broken down. The homologies of the legs and mouth parts should also be investigated. The gills will be found under the sides of the carapax. The relations and motions of the teeth found in the stomach will prove an interesting subject for study. Insects are dissected in much the same way as lobsters, but from their smaller size require more delicate manipulation.

Frogs are one of the best of vertebrates for study. In investigating the visceral anatomy they should be killed by *pithing;* a needle is forced into the spinal canal at the base of the skull and forced down the canal and also into the brain. The frog is now opened in the median ventral line and the parts carefully dissected out. The heart will continue beating for a considerable time after the animal has been killed. The brain is studied by opening the skull from the top. The muscles are easily dissected, their origin and insertion readily seen and the results produced by each one, readily understood. Fishes and many mammals are generally opened from the side. The left side is the one usually chosen, the head being directed toward the left hand of the operator.

The amount of time which can be advantageously spent on a single form or even on a single specimen is very great. In studying the anatomy of any form there should be no haste. Not a single cut should be made until the student realizes just why and what results will follow. It is far better to know the structure of one form well than to have a superficial and very vague idea of a dozen or more forms.

INJECTING.

The circulatory system is best studied in injected specimens. These are prepared as follows: an artery or other vessel is exposed and opened and in the opening the nozzle of the injecting apparatus is inserted. Usually an injecting syringe is use. This is a metal instrument, closely resembling the ordinary "surgeon's syringe," provided with nozzles of various sizes. Sometimes instead of a syringe an apparatus is used in which the weight of water or mercury is employed to force in the injection. This has the advantage of affording a steadier pressure then can be obtained in the ordinary manner. Fig. 37 represents this apparatus. Three bottles are required, each of which is corked with a stopper through which two glass tubes pass. One of each pair of tubes goes to the bottom of the bottle while the other merely passes through the corks. One bottle (b) is filled with water and is suspended by a string (a) passing over a pulley by which its height may be regulated. This bottle is connected with the second (d) by a rubber pipe attached to the long glass tube in each. This in turn is connected with the third by a second rubber pipe attached to the short glass tube and from this bottle runs a rubber pipe bearing the glass injecting nozzle. The methods of use are as follows: the tube connecting

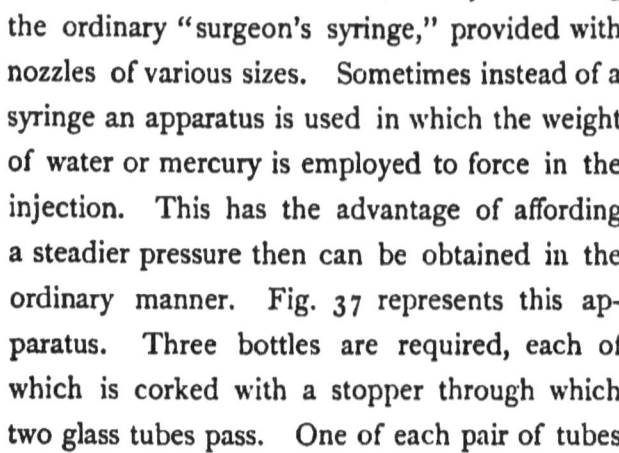

FIG. 37.

the first two bottles is filled and acts as a siphon and thus conveys the water to the second bottle creating there a pressure which in turn is communicated to the bottle, e, which contains the injecting material and which is thus forced out into the subject. By elevating or depressing the bottle b any desired pressure may be obtained.

With the injecting apparatus a colored fluid is forced into the vessels, rendering it very easy to trace them in dissecting. In some cases a saturated solution of prussian blue is sufficient for this purpose but it is better to employ albumen or gelatine as a basis.

Common gelatine is melted over a slow fire with the aid of water, in the proportions necessary to make a stiff jelly when cold. These proportions are usually given on the outside of the package. When melted, the coloring matter is stirred in. This may be an aqueous solution of carmine or prussian blue. Vermilion and yellow ochre are sometimes used but from their weight they are not readily held in suspension.

In injecting with gelatine the object must be first warmed and then kept in warm water while the operation is being performed, otherwise the jelly will set before the fine vessels are filled. To ensure success the specimen injected should be fresh, not alcoholic. When injected, the vessels should be tied and the specimen put away to cool so that the jelly may set. It is well to place it immediately in alcohol as that reagent facilitates the process by the extraction of water.

The place for the introduction of the injecting fluid varies with the form operated upon, as well as the system to be filled. In Vertebrates the arterial system may be injected from the aorta. In Crustacea a small hole should be bored through the carapax immediately over the heart and into this the nozzle of the syringe should be inserted, taking care that the tip enters, but does not pass through the heart. Injecting mollusks is rather more difficult, the injection should be forced into the heart, or into the vessel at the base of the gills; while starfishes are most easily injected by cutting off an arm and inserting the nozzle into the tube found on the lower surface. Insects and worms are only injected with very delicate manipulation. The method just described injects only the arterial system. To fill the veins is not so easy. It is best accomplished by forcing the injection into some distal portion and allowing it to find its way back to the heart. Thus the web of a frog's foot, the claw of a lobster and the foot of a mollusk are the best places for those forms. Great care must be exercised in performing an injection that the pressure employed is not sufficient to rupture the vessels. The more recently life is extinct the stronger the vessels are.

Besides injecting colored material to aid in the demonstration of the circulating system, various preservations are sometimes injected into the arteries, alimentary canal and body cavity to aid in keeping the specimens. Herr Wickerscheimer recommends his fluid for this purpose.

SECTION CUTTING.

It is necessary in making microscopic examinations of structures and tissues to have them thin enough to be transparent, or very translucent, otherwise they cannot be well illuminated. There are two ways of accomplishing this: by teasing and by cutting sections.

The easier method is by teasing. To do this a portion of the object is placed on a slide with the addition of a few drops of water, alcohol, glycerine or neutral salt solution according to the nature of the specimens and the objects or portions which it is desired to see. Then with two dissecting needles the tissue is teased or pulled into shreds, and then examined under the microscope, or mounted permanently after any desired method.

Teasing produces good results in fibrous tissue where it is desired to isolate the fibres, as in nerves, muscles, connective tissue, etc. At other times it does not work so well.

Section cutting is, however, the most universal and in the majority of cases the best method of preparing substances for examination, but at the same time it requires more time to accomplish. The various processes can conveniently be taken up in the following order: hardening and decalcifying, embedding, cutting and freeing from the embedding material.

Fresh tissues are generally either too soft, or in the case of bones, teeth, scales, and shell too hard to admit of being readily cut and hence certain steps must be taken to prepare them for the razor.

Suppose we have an object, an embryo tadpole for instance, of which we desire to obtain sections. This in its natural condition would be far too soft and must be hardened. This hardening may be accomplished in various ways. The most common method is first to place it for a few hours in weak alcohol (about 40 per cent.); it is then transferred to stronger (say 60 per cent.) spirit and after a short time is placed in strong alcohol (90 to 95 per cent.). The object after a day or two in this will be found to be much harder, and to possess a consistence fitting it for the use of the razor. The object of the successive uses of spirit of increasing strength is to prevent that contraction and distortion of the object which would occur were it placed at once in the strongest alcohol.

Müller's fluid is also extensively used for hardening objects, as are also chromic acid and Kleinenberg's picric acid. The *modus operandi* is essentially the same with either. The specimen is placed in a large quantity of the solution (Müller's fluid as directed on p. 138, or chromic acid $\frac{1}{2}$ and $\frac{1}{4}$ per cent.) and after a day or two is transferred to alcohol. These solutions must not be too strong nor must the specimens be kept too long in them, else they will become so brittle as to crumble under the section knife, rendering it impossible to obtain thin sections. A little experience will enable one to estimate the proper time for various tissues.

Osmic acid (one to one-tenth per cent. solution) is also very useful for hardening and at the same time it stains the section more or less darkly from a gray to a black. As noted on another page, it is selective in its staining, affecting nerves and fatty tissues more strongly than other tissues. The object is placed in the solution a varying number of hours according to the tissue, and then is washed thoroughly with distilled water and transferred to alcohol.

Other methods advocated by some students consist of the use of bichromate of potash, and among the older workers corrosive sublimate, but the foregoing answer all practical purposes.

When there are bone or lime salts in the tissues, chromic acid is the most useful reagent. It serves at the same time to harden the soft portions and to decalcify and thus soften the hard. The object must be placed in a large quantity of the fluid of a greater strength (one to two per cent.) and the acid should be frequently changed until all lime salts have disappeared. Of course, with this increase in strength of acid and the length of time of immersion, one runs a risk of the other portion becoming brittle, but this cannot well be avoided. In some cases it is necessary to use dilute hydrochloric acid in place of the chromic. This should rarely be used of greater strength than one per cent. After the substance is thoroughly decalcified it is transferred to alcohol.

Frequently specimens contain such a large amount of pigment matter as to render the thinnest section opaque and to utterly obscure all cell limits. In such cases it is necessary to immerse the tissues in 25 per cent. nitric acid, and to watch closely until the color disappears. This usually takes some hours, and the sections cut from such material are not very satisfactory but are the best that can be obtained.

The process of embedding comes next in order. The substances used are many, the most common being pure paraffine, a mixture of paraffine and oil or tallow, wax and tallow, transparent soap, gum arabic, and glycerine jelly. With most substances paraffine, without the admixture of anything, gives the best results, though many advise the addition of a fourth to a half of paraffine oil, or lard, or tallow, to render it softer.

It will be seen that in the above hardening processes the specimen was left in strong alcohol. This fits it for the next step, which is to soak it for a while (say half an hour) in spirits of turpentine. While the specimen is soaking the paraffine should be melted in a water bath (or in its absence a sand bath may suffice), over a spirit lamp or gas jet. A small portion of the paraffine should always be allowed to remain unmelted as thus the remainder will not acquire too high a temperature. When melted a portion of the paraffine is poured into a paper tray covering the bottom to the depth of an inch, and just allowed to "set." The object is then removed from the turpentine, the superfluous spirit being removed by blotting paper, and next placed on the surface of the paraffine in the tray and completely covered by more of the melted paraffine. When cold it is ready for cutting. The object when placed in the tray should be in such a position that the sections may be cut in the desired plane, and note should be taken of its position, as after the paraffine becomes hard this is difficult to ascertain.

In case the specimen to be embedded contains cavities, pains should be taken to fill these with paraffine. The usual method of doing this is to transfer the object directly from the turpentine to a mixture of half turpentine and half paraffine which is kept just melted. After a few minutes' immersion in this mixture it is transferred to the tray and the process completed as before.

A convenient tray may be made from common writing paper by taking a piece of proper proportions to the object to be embedded, longer than wide and folded on the lines shown in the accompanying figure. This is then made into a tray, the diagonals coming on the outside of the ends and then the portions which project on each of the shorter sides are folded down, thus holding the whole securely.

In case a section cutter with a hollow tube (*e. g.*, the Sterling microtome) be employed, it is better to embed directly in the tube, the process being essentially the same.

When soap is used for an embedding medium, the object is soaked in water instead of turpentine and the soap is melted with the addition of a slight amount of the same fluid. Otherwise the process is the same as before.

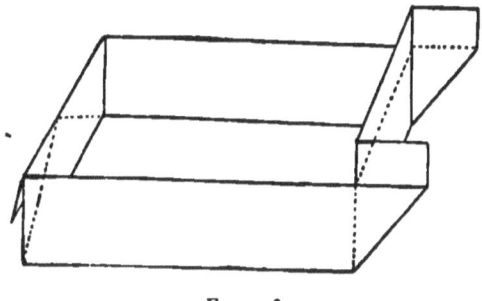

FIG. 38.

The foregoing methods both require heat and with many tissues as nerves, this produces an injurious effect. To avoid this gum arabic is employed. The specimen is washed in water and then placed in a thick mucilage of gum arabic and water, and the whole immersed in strong alcohol. The alcohol extracts the water and thus coagulates and hardens the whole.

Glycerine jelly requires heat but otherwise is used the same as gum arabic. Mr. Moseley found it very useful in studying the structure of *Millepora* and allied corals.

Elder pith is used by a few as an embedding material for some animal tissues, but I have had no experience with it. The object, surrounded with pith, is placed in the tube of the microtome and then the pith is wetted which causes it to swell and hold the whole firmly.

Of late years freezing has been a favorite method with those objects which are injured by heat. In this process the specimen is immersed in a mucilage of gum arabic in the tube of the microtome and frozen by an ether spray, or by having the tube surrounded by a tank in which is kept a freezing mixture of salt and ice.

Having embedded our object the next thing in order is to cut the sections. The first thing necessary is a knife. The most usual form is an ordinary razor, one side of which is ground flat. It is well to have the other side concave as thus a thinner edge is procured and the razor acts less like a wedge. At other times a knife made especially for the purpose is used. The knife, or razor, should be kept very sharp, and should have a perfect edge, free from any nicks, even if too small to be seen by the naked eye, as these will catch and ruin the section.

With practice good sections can be readily cut free hand, and it is always desirable that the student should be able to produce good results in this manner, whether he possess a microtome or not. In cutting sections free hand the bit of paraffine is held in the left hand and the razor either drawn towards or pushed from the operator with the right. Some work one way and some the other; the razor should not be pushed straight through but should have a drawing stroke,

When about to begin cutting the sections, the paraffine mass should be removed from the tray and trimmed to a convenient size, leaving more material behind the object than in front of it. Slices should then be carefully taken from above the object until it is reached and then even greater care should be taken. In case a mass of tissue is being cut it should be pared down until a good surface is reached, the slices taken off being rejected. When an embryo is being cut every slice, whether perfect or not, should be preserved. Always, when cutting sections from paraffine embeddings, the upper surface of the razor should be flooded with strong alcohol; when soap is used water replaces the alcohol and with glycerine jelly glycerine is useful. The object of this is to float the object up and prevent its sticking to the razor and thus becoming torn. It is convenient to have a shallow tank before the worker filled with alcohol or water, into which the razor with the section is dipped, the section being floated off and the razor wetted for the next section at the same time. When a sufficient number of sections have been cut from a paraffine embedding, the embedded material may be sealed up by placing a drop or two of melted paraffine on the cut end and the whole then labelled and put aside for future sections. A specimen thus embedded will keep for months without injury and may be cut from at any time.

FIG. 39.

After cutting, the sections are to be freed from their embedding material. In the case of paraffine this is accomplished by immersion in turpentine. When soap, or gum, or jelly is used, water will accomplish this. After being freed, they may be kept in alcohol or mounted as desired. In handling sections the greatest care should be exercised. A very convenient instrument is a section lifter, consisting of a thin sheet of metal attached at an angle to a handle. This is passed under a section floating in the liquid which is gently lifted and floated off in the desired place.

In case it be desired to keep the sections in consecutive order, each as cut must be transferred to its proper receptacle and properly labelled.

To aid in cutting sections mechanical appliances have been invented. These are known as microtomes or section cutters. Of these many forms have been in use, the best and most common being those described below.

The simplest form is the Sterling microtome (so called from its inventor). This consists of a tube in which moves a plug, regulated by a screw with a large graduated head. The other end of the tube bears a large brass or glass plate over the surface of which the razor passes.

The method of using is simple : the embedded material occupies the tube resting on the plug, a slight turn of the screw moves the whole forward, a slice is taken off with the razor, and the process is repeated.

By knowing the number of threads to the inch of the screw, and the fraction of a turn which it made in cutting each section, the thickness is an easy matter to ascertain. This section cutter (as in fact all others) produces sections with parallel surfaces, a rather difficult thing to obtain by cutting free hand.

Some in using this microtome prefer to hold it in the hand, others fasten it to the table and thus have both hands free for work.

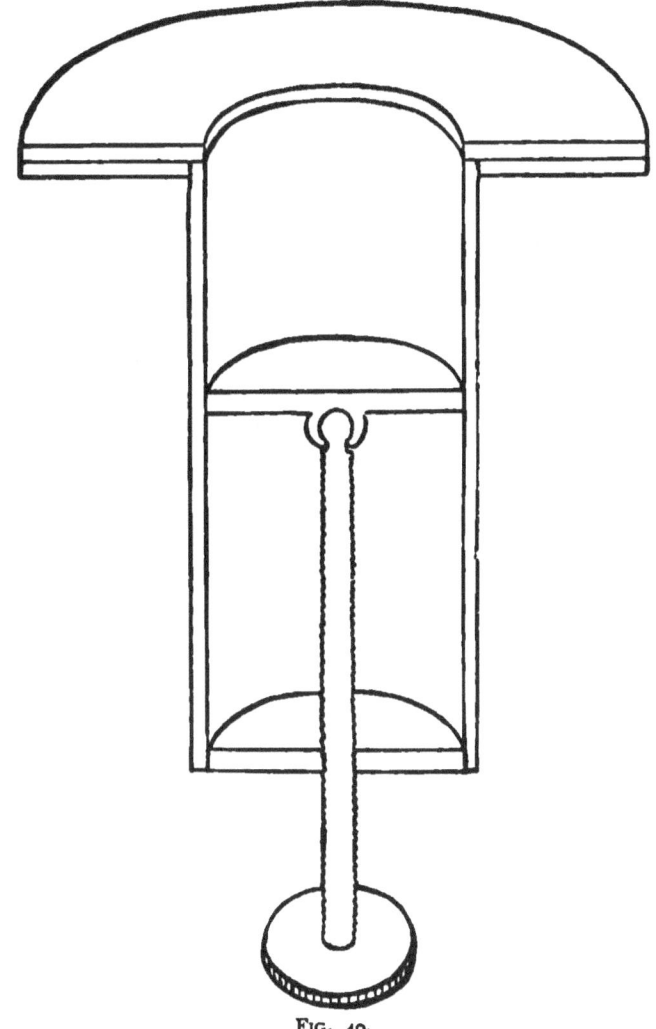

FIG. 40.

The next form to be described is the "Sledge Microtome" which was introduced to the American public by Dr. C. S. Minot. It is probably the best of these instruments.

It consists of a heavy oblong base about four by ten inches and about a quarter of an inch in thickness. Arising from this base is an upright plate; on either side of these are "ways," those on one side being horizontal and on the other slightly inclined. On the horizontal ones a carrier travels holding the knife, while on the other a second carrier is borne, moved by a screw. This second carrier holds the embedded specimen. Its method of operation is as follows: the embedded material is placed in the carrier which is moved forward and upward until it is high enough to cut. Then the knife is slowly but steadily drawn across the end of the embedded object. The knife is then returned, the screw turned the fraction of a revolution, and then another section cut. Of course the knife is to be constantly wet with alcohol. A valuable addition to this instrument may be made by having the object carrier mounted on a universal joint, thus permitting any desired inclination. Some sections of this model are made of wood and are good for nothing.

Dr. Seiler's microtome is well shown in the figure and needs no further description. It possesses this advantage over all others, that the knife has a drawing motion.

Freezing microtomes are sometimes employed. These are generally of the Sterling pattern and have an attachment by which a considerable degree of cold may be obtained either by ice and salt or by ether spray.

BOOKS FOR THE LABORATORY. 133

Balfour, F. M. Treatise on Comparative Embryology. 2 vols. (1 pub.), 8vo. London, 1880.

Beale, L. S. How to work with the Microscope. 5th edit. 8vo. London, 1880.

Brooks, W. K. Handbook of Invertebrate Zoöl. Boston, 1882.

Burden, Sanderson. Handbook for the Physiological Laboratory. 2 vols. 8vo. London, 1877.

Foster and Balfour. Elements of Embryology. 12mo. London, 1874.

Foster, M. Text Book of Physiology. 12mo. London, 1877.

Clark, H. J. Mind in Nature. 8vo. N. Y., 1865.

Gegenbaur, C. Elements of Comparative Anatomy. 8vo. London, 1878.

Huxley, T. H. Manual of the Anatomy of Vertebrated Animals. 12mo. London, 1871.

Huxley, T H. Manual of the Anatomy of Invertebrated Animals. 12mo. N. Y., 1878.

Huxley & Martin. Elementary Biology. 12mo. N. Y., 1876.

Milne-Edwards, H. Leçons sur la Physiologie et l'Anatomie comparée de l'homme et des Animaux. 13 vols., 8vo. Paris, 1857-79.

Owen, Richard. The Anatomy of Vertebrates. 3 vols., 8vo. London, 1868.

Ranvier. Traité d'Histologie. 8vo. Paris, 1875. 5 parts pub.

Rolliston, G. Forms of Animal Life 8vo Oxford, 1870.

Schäfer, E. A. A Course of Practical Histology. 8vo. London, 1877.

Siebold & Stannius. Comparative Anatomy (the Invertebrata translated by Burnett [Boston, 1851], the rest in German and French editions.)

Stricker S. Handbook of Human and Comparative Histology. 8vo. N. Y., 1872.

Tulk & Henfrey. Anatomical Manipulation. 12mo. Lond 1844.

CHAPTER VII.

RECIPES, FORMULÆ AND USEFUL HINTS.

Ammonia carmine.

Pulverized carmine 15 grains.
Aqua ammonia (strong) 40 drops.
Absolute alcohol 1 oz.
Glycerine 2 oz.
Distilled water 2 oz.

Put the carmine in a test tube. Pour in the ammonia, boil a few seconds and set the whole aside for a day to allow the superfluous ammonia to evaporate. Add the mixed glycerine and water, then the alcohol and filter.

Borax carmine.

Pulverized carmine 15 grains.
Saturated aqueous solution of borax 15 fl. dr.
Mix and add absolute alcohol 15 fl. dr.

Filter and collect the crystals when dry. Dissolve nine grains of the crystals in one ounce of distilled water.

Picrocarminate of ammonia (Picrocarmine).

Make a strong solution of carmine in ammonia and water, and a saturated solution of picric acid in water. Mix the acid solution with the carmine until the ammonia is just neutralized. Then filter.

Hæmatoxylin.

Rub together in a mortar one part extract of logwood and three parts pulverized alum (by measure) and afterward add enough water to dissolve only a portion of the resulting powder. Let the solution stand several days until it becomes saturated and assumes a dark violet color. If still too red add more alum. Then filter and dilute with one-fourth of seventy-five per cent. alcohol.

Hæmatoxylin (No. 2).

Ground Campeachy wood	1 oz.
Pulverized alum	2 oz.

Mix and triturate in a mortar for twenty minutes; then add two ounces of hot distilled water, and let the whole stand a couple of days. Filter and add to each ounce a quarter of an ounce of ordinary alcohol. After standing twenty-four hours more, filter again to remove the precipitated alum. This will keep two months in a well stoppered vial.

Hæmatoxylin. (No. 3).

Hæmatoxylin crystals	35 parts.
Absolute alcohol	1000 parts.
Water (distilled)	3000 parts.
Alum	10 parts.

Dissolve the hæmatoxylin in the alcohol, the alum in the water and mix. The mixture is purple at first, but gradually turns blue. It can however be used at once after filtering.

Hæmatoxylin (Kleinenberg's Method).

Make a saturated solution of crystallized chloride of calcium in 70 per cent. alcohol, and add alum until no more will be dissolved. This is the first solution; the second is a saturated solution of alum in 70 per cent. alcohol. Mix these two solutions in the proportions one of the first to eight of the second. Then to the mixture add a few drops of a saturated solution of crystallized hæmatoxylin in absolute alcohol.

Frey's Fuschine solution.

Crystallized fuschine	1 centigramme.
Absolute alcohol	15 – 20 drops.
Distilled water	15 cu. centimetres.

This, though a useful staining medium, possesses the disadvantage that it cannot be used where the tissue is to be mounted in balsam.

RECIPES, FORMULÆ AND USEFUL HINTS. 137

Eosin and fuschine (aniline colors) are used in a two per cent. aqueous solution but have a tendency to stain all parts equally. The method of operation is similar to that with carmine.

Nitrate of silver is used for differentiating the walls of cells. The object is placed for a few moments in a solution made of nitrate of silver, one part, *distilled* water, 500 parts, then washed in distilled water and exposed to the direct rays of the sun. It is then mounted as may be desired and shows the cell walls black, and in many cases this is the only way in which they can be seen at all.

Osmic acid, in one-tenth per cent. solution, is useful fo studying nerves and fatty tissues as it stains them black. Great care should be taken in using osmic acid as it is one of the most poisonous substances known, and its vapor badly affects the eyes and nasal passages.

Chloride of gold is also of value in studying the histology of the nerves. Sections are first placed from five to seven minutes in a one half per cent. solution of the chloride, then soaked in distilled water several hours, next placed in water acidulated with lactic acid to reduce the chloride, and after the proper color (a dark purple) is obtained, the specimens are washed with distilled water and soaked in alcohol and mounted as may be desired.

Moleschott's acetic acid and alcohol (strong).
    Strong acetic acid (sp. gr. 1.070)   1 part.
    Strong alcohol   1 part.
    Distilled water   2 parts.

This hardens many organs, makes connective tissue transparent and renders albumen distinct.

Moleschott's acetic acid (weak).
    Acetic acid (sp. gr. 1.070)   1 part.
    Alcohol   25 parts.
    Distilled water   50 parts.

This is better for delicate tissues than the strong. Besides rendering connective tissue transparent, acetic acid renders the nuclei of cells more plainly visible. When a one per cent. solution of acetic acid is mixed with about one-third its bulk of ordinary hæmatoxylin solution, the connective tissue corpuscles are stained a beautiful purple.

Müller's Fluid.

Sulphate of soda	1 part.
Bichromate of potash	2 parts.
Water	100 parts.

Müller's fluid for hardening the retina.

Bichromate of potash	2½ grammes.
Sulphate of soda	1 gramme.
Distilled water	100 grammes.

Besides its hardening properties, this is useful for preserving glands, mucous membranes and ciliated cells.

Bichromate of potash for macerating specimens is used in a ¼ per cent. aqueous solution.

Iodized serum.

Take the amniotic fluid of a cow or other mammal, add a crystal or two of iodine and allow it to stand a few days, with frequent agitation. Then filter. Any other serous fluid may be used. Iodized serum seems to macerate tissues and at the same time to preserve the original form of the histological elements. The iodine tends to prevent putrefaction and at the same time renders the elements of the tissue firmer.

Artificial iodized serum.

Take one ounce of the white of an egg; pick out the chalazeæ and mix with nine ounces of water and fifty grains of common salt. Then add several crystals of iodine.

Neutral salt solution is a ½ per cent. solution of common salt in distilled water. It is useful in studying living tissues as the cells do not exhibit such marked changes as when pure water is used.

Glycerine jelly.

Cooper's gelatine	1 oz.
Best glycerine	6 oz.
Carbolic acid	20 drops.

Soak the gelatine over night in water; in the morning add the swelled gelatine to the glycerine and carbolic acid heated to about 200° Fahr. in a water bath. Continue the heating several hours until the water is all expelled. Then filter and bottle. The filtering is difficult and can only be accomplished by the aid of heat.

Glycerine jelly.

The original method of making this is as follows: Take any quantity of gelatine and let it soak several hours in cold water. Pour off the superfluous water and melt the soaked gelatine by the aid of heat. To each ounce of the fluid gelatine add one drachm of alcohol and mix well. Then add a fluid drachm of the white of an egg and mix well while the gelatine is cool but still fluid. Now boil until the albumen coagulates and the gelatine is quite clear. Filter through fine flannel, and to each fluid ounce of the clarified gelatine add six fluid drachms of pure glycerine (Price's is the best) and mix well.

Glycerine and gum.

Dissolve two parts by weight of gum arabic in two parts of cold water and add one part of glycerine. Mix well but use no heat, and strain. Keep in a tightly stoppered vial. This medium has the advantage in mounting, that no heat is required, while it becomes solid in a short time after mounting.

Dr. Lang's method of studying nervous histology of the Turbellaria.

50 parts   1 per cent. solution of Picrocarminate of ammonia.
50 parts   2 per cent. aqueous solution of eosin.

Objects are hardened in alcohol and placed in this solution one-half to four days. The picric acid is then extracted by 70 per cent. alcohol, and the specimens washed with 90 per cent. and absolute alcohol as long as any eosin is dissolved. In embedding in paraffine a copious use of creasote is recommended. This produces in sections of *Dendrocœla*, carmine red nuclei and nucleoli, glands, adipose tissue, while all other parts are eosin red.

Dr. Treub, in studying the *nuclei of plants*, first killed the cells by absolute alcohol; then placed the tissues in 1 per. cent solution of picro-carminate from four to twelve hours. They were next shaken in distilled water to remove the picric acid and placed in glycerine and water which is gradually replaced by pure glycerine containing 1 per cent. of formic acid. By this proces the nuclei are stained bright red, the rest of the cell being uncolored.

Goadby's solution for preserving specimens.

No. 1. Bay salt   4 oz.
        Alum   2 oz.
        Corrosive sublimate 2 grs.
        Rain water   1 qt.

No. 2. Salt.   ½ lb.
        White arsenic   20 grs.
        Corrosive sublimate 2 grs.
        Boiling rain-water 1 qt.

Arsenical soap.
   BECŒURS.

Camphor   5 oz.
White arsenic   2 lbs.
White soap.   2 lbs.
Salts of tartar   12 oz.
Powdered chalk   4 oz.

The soap is melted with a little water over the fire, and the chalk and tartar added. It is then removed from the fire and the arsenic, and lastly the camphor mixed with a little alcohol is stirred in. The paste is then packed in jars and labelled.

Arsenical soap.
   SWAINSON'S.

White arsenic   1 oz.
White soap   1 oz.
Carbonate of potash   1 dr.
Distilled water   6 dr.
Camphor   2 dr.

This composition is formed in cakes like ordinary soap.

Arsenical soap.

SIMON'S.

Soap	1¼ lb.
Alum	8 oz.
Carbonate of potash	4 oz.
Common salt	4 oz.
Powdered chalk	8 oz.
Powdered camphor	2 oz.
Water	1 pt.

Melted by heat, the camphor being added last.

Arsenical soap.

LAURENT'S.

Arsenite of potash	2 dr.
Alum	2 dr.
Powdered camphor	2 dr.
White soap	½ oz.
Alcohol	6 oz.

The first two placed in a bottle and the alcohol poured over them. When dissolved, the other ingredients are added. This composition requires to be tightly corked.

Bullock's arsenical powder.

White arsenic	1 lb.
Burnt alum	1 lb.
Tanner's bark	2 lbs.
Camphor	½ oz.
Musk	½ oz.

The first three to be firmly powdered and passed through a sieve, the others then to be added and the whole thoroughly mixed.

Corrosive sublimate solution.

Dissolve one ounce of corrosive sublimate in one quart of alcohol in a glass vessel. This solution is to be applied with a string-wound brush as the presence of metal will produce a discoloration.

"*Sugar*" *for moths.*

Ale	½ pint.
Honey	½ lb.
Sugar	¼ lb.
Rum	1 oz.
Oil of bitter almond	5 drops.

The ale is heated and the sugar and honey added. When cold the rum and oil of almond, having been previously mixed are poured in and the whole thoroughly stirred.

"*Sugar*" *for moths.* No. 2.

A thick sugar made of brown sugar with a small quantity of rum.

Dr. Leconte's poison for insects.

Saturated alcoholic solution of arsenic	8 fl. oz.
Strychnine	12 grs.
Crystallized carbolic acid	1 dr.

Heavy benzine and alcohol enough to make one quart.

Heavy benzine should be used (about 10-12 oz.), as lighter will not mix with alcohol. The benzine should be tested for grease, by moistening paper with it. If all greasy appearance does not disappear on drying it should be rejected. This poison is to be applied to the insects in the cabinet by an atomizer.

A good mucilage.

Take equal parts of gum arabic and gum tragacanth, swell in water and then dissolve by means of heat, then add a few drops of carbolic acid and a few of glycerine. The carbolic acid prevents fermentation or mould, the glycerine keeps it from cracking or scaling off when dry.

Thick flour paste added to common glue adheres well to glass as also does the mucilage made of gum arabic and gum tragacanth.

"*Electrical Cement.*"

Melt together ten oz. of resin, two oz. of beeswax, two oz. of red ochre, and add a teaspoonful of plaster of Paris. This is used hot for cementing brass or wood to glass.

Peron's Luting.
Common resin
Red ochre
Yellow wax
Oil of turpentine

First melt the wax, then add the resin, next stir in the ochre and lastly the turpentine. According as the ochre or other ingredients predominate, the luting will be brittle or elastic. Great care should be taken that the mixture does not take fire and the vessel used should be capable of containing at least three times the quantity made at one time.

Grafting wax.

Melt together eight oz. resin, three of beeswax and one of lard. Run in sticks. It improves with age.

Black ink.

A black ink is frequently desirable and is almost impossible to obtain in the stores. A good ink may be made by boiling eight oz. of powdered nut galls and one oz. extract of logwood in three quarts of water for an hour. Strain and add four oz. of copperas (sulphate of iron), three oz. of gum arabic and one of blue vitriol (sulphate of copper); let it stand until dissolved and strain again. A few cloves will keep it from moulding.

Old alcohol which has been discolored by specimens can be cleaned by filtering through animal charcoal, but nothing will completely remove disagreeable odors though a redistillation will sometimes help it. After filtering, the spirit should be brought to a proper strength (to be ascertained by the hydrometer) by adding new alcohol.

To blacken brass.

It is occasionally desirable to blacken portions of instruments as stages of microscopes, etc. This may be done by cleaning the brass of all grease, then covering with a solution of nitrate of copper which on the application of heat turns the surface to a jet black. If desirable, it may then be lacquered by applying shellac varnish and heating slightly.

BIBLIOGRAPHY.

USEFUL WORKS OF REFERENCE.

The following list includes only such works as will aid the student in arranging and identifying his collections, all morphological papers being purposely omitted. The more useful of these are printed in full face type. While the list is far from perfect it is hoped that it will prove of use to the zoölogist.

The majority of the titles have been translated either in full or in abstract, but the language in which the article is written is indicated by the abbreviation following the title. For a more complete list of papers, students should refer to the various special bibliographies quoted. The catalogues published by Friedländer und Sohn of Berlin will also prove useful.

All Museums should possess Dictionaries of French, German, Latin, Swedish, Danish, Spanish, Italian, Portuguese and Norse languages ; as many important works are published in those tongues.

Other necessary works are :

Agassiz L., Nomenclator Zoölogicus, 4to, Solduri, 1846-48.

Marschall, A. D., Nomenclator Zoölogicus, 8vo, Wien, 1873 (a continuation of Agassiz's work).

A large Atlas of the world.

Lippincott's complete pronouncing Gazetteer, 2 vols., 8vo, Philadelphia, 1880.

Scudder, S. H., Catalogues of Scientific Serials of all countries, 8vo, Cambridge, 1879.

Royal Society's list of Scientific Papers, 8 vols, 4to, London, 1868.

CHAPTER VIII.

BIBLIOGRAPHY.

GENERAL ZOÖLOGY.

Bonaparte, C. S. Iconografia della Fauna Italica. 3 vols., fol. Rome, 1832-41. (Ital.)

Brehm, A. E. Thierleben (Animal life, a general account of the Animal Kingdom). 10 vols., 8vo. Leipzig, 1868-78. (Ger.)

Bronn und Gerstaecker. Klassen und Ordnung der Thierreich. 8vo. Many plates. Leipzig, 1863— (Ger.)

Cooper, Suckley and Gibbs. Zoölogy in vol. XII, Reports Pacific R. R. Survey. 4to. Washington, 1860. (Separate as Natural History of Washington Territory).

Cuvier Regne Animal. Edition by Audouin, Blanchard, Deshayes, Milne Edwards, Valenciennes, etc. 22 vols., 4to, 903 pls. Paris, 1849, *et seq.*

Donovan, E. Naturalists' Repository. 5 vols. London, 1834.

Eydoux et Gervais. Voyage of the Favorite. Paris, 1839. (Fr.)

Eydoux et Souleyet. Voyage of the Bonite. 2 vols., 8vo, and folio Atlas. Paris, 1841-52. (Fr.)

Fabricius, O Fauna Grœnlandica. 8vo. Leipzig, 1780.

Gay, C. History of Chili. 5 vols., 8vo, 4to plates. Paris, 1847. (Sp.) Zoölogy by Gay, Nicollet, Blanchard, etc.

(149)

- Gmelin. Systema Naturæ. 3 vols., 8vo. Leipzig, 1788-93. (Lat.)

Gray, J. E. Voyage of H. M. S. Sulphur. 4to. London, 1843-45. Illustrations of Indian Zoölogy. 20 pts., 4to. London, 1830-34.

Griffith and Henfrey. The Micrographic Dictionary. 8vo. London. 3 edit., 1871-75.

Guerin-Meneville. Iconographie du Regne Animal. 3 vols., 8vo. 450 pls. Paris, 1829-44. (Fr.)

Harlan, R., Medical and Physical Researches. 8vo. Philadelphia, 1835.

Van der Hoeven. Handbook of Zoölogy, translated by Wm. Clark. 2 vols., 8vo. London, 1856-58.

Jacquemont. Voyage to India. 4 vols., 4to. fol. Atlas, 229 pls. Paris, 1841-44. (Fr.)

Jardine, W. Naturalists' Library. 40 vols., 12mo. London, 1834-43.

Leach, W. E. Zoölogical Miscellany. 3 vols., 8vo. Many pls. London, 1814-17.

Lesson et Garnot. Voyage of the Coquille. 2 vols., 4to. 157 folio plates. Paris, 1826-30. (Fr.)

Linnæus, C. Systema Naturæ. Edit. x, 2 vols., 8vo. 1758-9. Edit. xii, 3 vols., 8vo. 1765. (Lat.)

Müller, O. F. Zoölogia Danica. 4 vols. Copenhagen, 1788-1806. (Lat.)

Nicholson, H. A. Advanced Textbook of Zoölogy. 8vo. London, 1870.

Packard, A. S., jr. Zoölogy. 12mo. N. Y., 1879.

Pagenstecher, A. General Zoölogy. 8vo. Berlin, 1875. (Ger.)

Pallas, P. S. Miscellanea Zoölogica, 4to. 1766. (Lat.)

Peters, Carus und Gerstaecker. Handbuch der Zoölogie. 2 vols., 8vo. Leipzig, 1863-68. (An indispensable work. Ger.)

Quoy et Gaimard. Voyage of the Astrolabe. 4 vols., 8vo. 192 folio plates. Paris, 1830-33. (Fr.)
Raman de la Sagra. History of Cuba, 13 vols., 8vo. 250 fol. pls. Madrid et Paris, 1849-61. (Fr. and Spa.)
Richardson, J. Fauna Boreali Americanæ. 4to. London, 1829. Report on North American Zoölogy (6th Rep. Brit. Assoc. Adv. Sci., 1737).
Record of Zoölogical Literature. 8vo. London, 1865. (yearly).
Risso. Natural History of Central Europe. 5 vols. 8vo. 1828. (Fr.)
Say, T. Zoölogy in Long's Expedition to Rocky Mts. 2 vols. fol. Leyden, 1839-45.
Schlegel, Müller et De Haan. Fauna of the Dutch East Indies. 8vo. Philadelphia, 1823.
Schmarda, L. K. Zoölogie. 2 vols., 8vo. Vienna, 1877-78. (Ger.)
Shaw, G. General Zoölogy. 14 vols., 8vo. London, 1809-26.
Shaw and Nodder. Naturalists' Miscellany. 24 vols., 8vo. 1000 pls. London, 1789-1814.
Siebold, Schlegel et De Haan. Fauna Japonica. 5 vols., fol. 1838-50. (Fr. and Lat.)
Sowerby, J. British Miscellany. 4 vols., 8vo. London, 1804-06.
Tenney, S. Manual of Zoölogy. 8vo. N. Y., 1874.
Turton, W. General System of Nature. 7 vols., 8vo. London, 1806.

VERTEBRATA.

Coues and Yarrow. Zoölogy. Vol. 5 of Reports of Wheeler's Survey. 4to. Washington, 1875.
Fitzinger. Atlas of Natural History of the Vertebrates. 4 pts., 4to. 474 pls. Vienna, 1856-64. (Ger.)

Gervais, P. New or rare animals of Castlenau's expedition to South America. Paris, 1865. (Fr.)

Godman, J. I. American Natural History. 3 vols., 8vo, 2nd edit. Phila., 1831.

Huxley and Hawkins Elementary Atlas of Comparative Osteology. London, 1864.

Jordan, D. S. Manual of the Vertebrata of the Northern United States, east of the Mississippi and north of North Carolina and Tennessee, exclusive of Marine species. 8vo. 2 edit. Chicago. 1878.

List of Vertebrates in the Gardens of the Zoölogical Society of London. 8vo. 6th edit., 1877.

Middendorff, A. T. von. Siberian Journeys. Vertebrata. 4to. St. Petersburg, 1853. (Ger.)

Owen, R. Odontography or Comparative Anatomy of the Teeth. 2 vols., 8vo. London, 1840-45. Description of the osteological series in the Museum of the Royal College of Surgeons of England. 2 vols. London, 1853.

Pander and d'Alton. Comparative Osteology. 13 pts. Rome, 1821-31. (Ger.)

Rüppell, E. Zoölogical atlas of a Journey in North Africa. 5 parts, fol. Franckfurt, 1826-31. (Ger.)

Wied, Neuwied, Prince Maxmillian. Illustrations of the natural History of Brazil. fol. Weimar, 1822-31. (Ger.)

MAMMALIA.

(For a more complete list see Coues and Allen, Monographs Rodentia.)

Allen, H. Monograph of the Bats of North America. (Smithsonian Misc. Collections, vol. vii). 8vo. Washington, 1874.

Allen, J. A. Catalogue of the Mammals of Massachusetts.

(Bulletin Museum Comp. Zoöl. i.) 8vo. Cambridge, 1869.

Allen, J. A. Catalogue of mammals and winter birds of Florida. (Bull. M. C. Z. ii.) 8vo. Cambridge, 1871.

Allen, J. A. The species of Bassaris. (Bulletin, U. S. Geol. Surv., v.) 8vo. Washington, 1879.

Allen and Bryant. On the Otariidæ or eared Seals of the North Pacific. (B. M. C. Z. ii.) 8vo. Cambridge, 1870.

Audubon and Bachman. The Viviparous Quadrupeds of North America. 3 vols., 8vo. New York, 1846-54.

Baird, S. F. Mammalia in Rep. U. S. Pacific R. R. Surveys, vols. viii-x. 4to. Washington, 1857-59.

Baird, S. F. Mammalia in Rep. U. S. and Mex. Boundary Survey. 4to. Washington, 1859. (The two above are bound together under the title, Mammals of North America. Washington, 1859).

Baird, S. F. Catalogue of North American Mammalia. 4to. Washington, 1857.

Bell, T. History of British Quadrupeds including the Cetacea. 1 vol., 8vo. London, 1837. Another edition edited by Thomes and Alston, 1874.

Buffon. Natural History of Apes and Quadrupeds. 15 vols., 8vo. Paris, 1800. (Fr.)

Burmeister, H. Systematic Review of Brazilian Animals. 1 vol., 8vo. Berlin, 1854. (Ger.)

Blainville, H. M. D. de. Osteography or Comparative Iconography of the Skeleton and Teeth of recent and fossil Mammalia. Text, 4 vols., 4to, plates. 4 vols., folio. Paris, 1839-64. (Fr.)

Blanford, W. T. Mammalia of Second Yarkand Mission, Calcutta, 1878.

Brandt. Contributions to a better knowledge of the Russian Mammalia. (Mem. Acad. St. Petersburg, ix, 1855.) 4to. (Ger.)

Cassin, J. Mammals and Birds of the U. S. Exploring Expedition. 4to. folio atlas. Phila., 1858.

Cope, E. D. Contributions to the History of the Cetacea.
Coues, E. Fur Bearing Animals. (Misc. Pub. U. S. Geol. Surv. 8vo. Washington, 1877.
Coues, E. Papers in Proceedings, Acad. Nat. Sci., Phil., Bulletin U. S. Geol. Surv, etc.
Coues and Allen. Monographs of the North American Rodentia. (Final Reports of U. S. Geol. Survey, vol. xi.) 4to. Washington, 1877.
DeKay, J. E. Zoölogy of New York. Pt. I, Mammalia. 4to. Albany, 1842.
Desmarest, A. G. Mammalogy, or description of the species of Mammals. 2 vols., 4to. Paris, 1820-22. Also in Encyclopedie Methodique, vol. 182. (Fr.)
Dobson, G. E. Catalogue of the Cheiroptera in the British Museum. 8vo. London, 1878.
Emmons, E. Report on the Quadrupeds of Massachusetts. 8vo. Cambridge, 1840.
Erxleben, J. C. System of the Animal Kingdom. Pt. I Mammalia. 8vo. Leipzig, 1777. (Lat.)
Eschricht, Reinhardt and Lilljeborg. Memoirs on the Cetacea. Edited by W. H. Flower (Ray Society.) Folio. London, 1866.
Fitzinger, L. J. Critical Review of the Cheiroptera. 8vo. Vienna, 1869-71. (Ger.)
Flower, W. H. Introduction to the Osteology of the Mammalia. 8vo. Lond., 1870. 2nd edit., Lond., 1877.
Geoffroy St. Hilaire, I. Description of new or little known monkeys. 3 parts, 4to. Paris, 1841-51. (Fr.)
Geoffroy. Mammals and Birds in Jacquemont's Voyage to the East Indies. 4to. Paris, 1842-3. (Fr.)
Geoffroy. Voyage of the Venus Around the World Paris, 1855. (Fr.)
Geoffroy St. Hilaire and Cuvier. Natural History of Mammals. 3 vols., folio. Paris, 1819-29. (Fr.)

Gervais, P. Natural History of Mammals. 2 vols., 8vo. Paris, 1854-55. (Fr.)

Giebel, C. G. Natural History of the Animal Kingdom. Vol. 1, Mammals. 1 vol. 4to. Leipzig, 1859. (Ger.)

Gray, J. E. **Lists and Catalogues of Mammalia in the British Museum.** (Several volumes under distinct titles.) 1843-74.

Guerin, R. Zoölogical and Paleontological studies of the Cetacea. 4t . Paris, 1874. (Fr.)

Harlan, R. Fauna Americana. Descriptions of the Mammalia of North America. 1 vol., 8vo. Philadelphia, 1825.

Horsfield, T. Zoological researches in Java. 4to. London, 1824.

Jardine, Sir W. Naturalists' Library. Mammalia (by Jardine, Waterhouse, Macgillivray and others). 13 vols., 8vo. Edinburg, 1833-42.

Jerdon, T. C. The Mammals of India. 8vo. London, 1874.

Kennerly, C. B. Mammals of Upper California in Pacific R. R. Survey, vol. x, 4to. Washington, 1859.

Locke. Mammals and Birds of the Scientific exploration of Algiers. 4 vols., 4to. Paris, 1867. (Fr.)

Milne Edwards, H. et A. Researches in the Natural History of the Mammalia. 1 vol., 4to. 105 Plates. Paris, 1868-74. (Fr.)

Peters, W. Mammalia of Travels in Mozambique. 1 vol., fol. Berlin, 1852. (Ger.)

Pucheran. Monograph of the Genus Cervus. 4to Paris, 1852.

Reichenbach, H. G. L. Complete Natural History of the Apes. 4to. Dresden, 1863. (Ger.)

Richardson, J. Quadrupeds of the Fauna of Northern America. 4to. London, 1829-36.

Richardson, J. Mammals of Beechey's Voyage. 4to. London, 1839.

Sagra and d'Orbigny. Mammals and Birds in de Sagra's History of Cuba. Paris, 1840. (Fr.)

Scammon, C. M. The Marine Mammals of the North Western Coast of North America. 8vo. San Francisco, 1873.

Schinz, H. R. Illustrations and Natural History of the Mammalia. 2 vol., fol. Zürich, 1827. (Ger.)

Schinz, H. R. Monographs of the Mammalia. 4to. Zürich, 1843-56. (Ger.)

Schlegel, H. Monograph of the Apes. Leyden, 1876. (Fr.)

Schlegel and Pollen. Investigation of the Mammals and Birds of Madagascar. Folio. Leyden, 1868. (Fr.)

Schreber, J. C. D. The Mammalia in pictures from Nature with descriptions continued by J. A. Wagner, 1775-92, 1840-55.

Siebold, Temminck and Schlegel. Mammalia of the Fauna Japonica. fol. Leyden, 1842. (Fr.)

Slack, J. H. Monograph of the Prehensile tailed Apes. (Proc. Acad. Nat. Sci., Phila., 1862). 8vo.

Temminck, C. J. Monograph of the Mammalia. 2 vols., 4to. Paris and Leyden, 1827-41. (Fr.)

Tschudi, J. J. Investigations of the Peruvian Fauna. Mammals and Reptiles. 4to. St. Gallen, 1844-46. (Ger.)

Waterhouse, G. R. Natural History of the Mammalia. 2 vols., 8vo. London, 1846-48.

BIRDS.

Allen, J. A. Catalogue of Mammals and winter Birds of Eastern Florida. (Bulletin Mus. Comp. Zoöl., ii, 1871.)

Audubon, J. J. Ornithological Biography. 5 vols., 8vo. Edinburg, 1831-39.

BIBLIOGRAPHY.   157

Audubon, J. J. Birds of America from original drawings. 5 vols., fol. London, 1827-38.

Audubon, J. J. Birds of America, from drawings made in the U. S. and their Territories. 7 vols., 8vo. N. Y., 1840-44.

Baird, S. F. Birds in Stansbury's Salt Lake. 8vo. 1853

Baird, S. F. Birds in Rep. Mexican Boundary Survey. 4to. Washington, 1859.

Baird, S. F. Catalogue of North American Birds. (Smithsonian Inst. Misc. Coll., ii, 1859).

Baird, S. F. Birds, vol. ix of Report Pacific Railroad Survey. 4to. Washington, 1859.

Baird, S. F. Review of American Birds in the Smithsonian Museum. (S. I. Misc. Coll. xii). 8vo Washington, 1864.

Baird, Brewer and Ridgway. A History of North American Birds. (Land Birds complete, 3 vols.)

Baird, Cassin and Lawrence. The Birds of North America. 4to. Philadelphia, 1860.

Bannister, W. H. Classification of the American Anserinæ (Proc. Phila. Acad. 1870).

Barrows, W. B. Catalogue of the Alcidæ in the Collection of the Boston Soc. Nat. Hist. (Proc. Bost. Soc. xix, 877).

Blanford, W. T. Monograph of the genus Saxicola. London, 1870

Blyth, E. Catalogue of Birds in the Museum of the Asiatic Society. 8vo. Calcutta, 1849.·

Bonaparte, C. S. American Ornithology or Nat. Hist. of Birds of North America, not given by Wilson. 4 vols , 4to. N. Y., 1825-33.

Bonaparte, C. S. Genera of N. A. Birds and a synopsis of those found in the U. S. (Annals N. Y. Lyceum Nat. Hist ii, 1826.) Also separate, 1828.

Bonaparte, C. S. Monographs of the Shrikes, Parroquets, Birds of Prey and Gulls. Revue et Magazin de Zoölogie, 1853–55. (Fr.)

Bonaparte C. S. Conspectus of the Genera of Birds edited by Finsch. Leyden, 1850-65. 2 vols., 8vo. (Lat.)

Bonaparte, C. S. Iconography of the Pigeons. fol. Paris, 1857. (Fr.)

Bonaparte and Schlegel. Monograph of the Cross Bills. 4to. Leyden, 1850. (Ger.)

Boucard, A. Catalogue of all described Birds. 8vo. London, 1876.

Brehm, C. L. Handbook of the Birds of Europe. 8vo. Ilmenau, 1831. (Ger.)

Brewer, T. W. North American Oölogy. Raptores and Fissirostres. (Smithsonian Contrib. to Knowl. xi). 4to. Washington, 1857.

Buffon. Natural History of Birds. 10 vols., 4to. 1008 Pls. Paris, 1770-86. (Fr.)

Burmeister, H. Systematic Review of Brazilian Animals. 8vo. Berlin, 1854. (Ger.)

Cassin, J. Mammals and Birds of the U. S. Exploring Expedition. 4to. atlas folio. Phila., 1858.

Cassin, J. Illustrations of the Birds of California, Texas, Oregon, British and Russian America. (1st series 50 pls). 4to. Phila., 1853-55.

Cooper and Baird. Ornithology of California. vol. 1. Land Birds. 4to. (Pub. Cal. Geol. Survey). San Francisco, 1873.

Costa, O. G. Ornithology of the Kingdom of Naples. 4to. Naples, 1857. (Ital.)

Coues, E. Monograph of the Alcidæ. (Proc. Phila. Acad. 1868).

Coues, E. Key to North American Birds. 8vo. Salem, 1872.

Coues, E. Materials for a Monograph of the Spheniscidæ. (Proc. Phila. Acad., 1872).

Coues, E. Revision of the species of Mylarchus. (Proc. Phila. Acad., 1872.

Coues, E. Field Ornithology with a check list of North American Birds. 8vo. Salem, 1874.

Coues, E. Birds of the North West. (Hayden's Survey). 8vo. Washington, 1874.

Coues, E. Birds of the Colorado Valley. (Hayden's Survey). 8vo. Washington, 1878.

Daudin, F. M. Ornithological Treatise. 2 vols., 4to. Paris, 1800. (Fr.)

Degland et Gerbe. European Ornithology. 2 vols., 8vo. 2 edit. Paris, 1867. (Fr.)

DeKay, J. E. N. Y. Fauna, Pt. ii, Birds. 4to. Albany, 1844.

Des Murs, P. O. Iconographic Ornithology. 4to. Paris, 1845-49. (Fr.)

Donovan. Natural History of British Birds. 11 vols., 8vo. London, 1794-1818.

Dresser and Sharpe. History of the Birds of Europe. 74 pts. pub. 4to, 480 pls. London, 1871-69 et seq.

Elliot, D. G. Monograph of the Psittidæ. fol. N. Y., 1861-62.

Elliot, D. G. Monograph of the Tetraonidæ. fol. N. Y., 1865.

- Elliot, D. G. Monograph of the Phasianidæ. 2 vols., fol. London, 1872.

Elliot, D. G. Monograph of the Birds of Paradise. fol. London, 1875.

Elliot, D. G. Monograph of the Horn Bills. Folio. London, 1876 et seq.

Eyton, T. C. Monograph of the Anatidæ. 4to. London, 1838.

Fritsch, A. Natural History of European Birds.
8vo, fol., atlas. Prag, 1858-1871. (Ger.)

Geoffroy St. Hilaire. Mammals and Birds in Jacquemont's Voyage to the East Indies. 4to. Paris, 1842-43. (Fr.)

Gould, J. Birds in the Voyage of the Beagle. London, 1841-44.

Gould, J. Birds of Europe. 5 vols. folio. London, 1832-37.

Gould, J. Synopsis of the Birds of Australia. Fol. London, 1834.

Gould, J. Monograph of the Ramphastidæ. Folio. London.

Gould, J. Monograph of the Trogonidæ. Folio. London, 1838.

Gould, J. Monograph of the Partridges of America. Folio. London, 1850.

Gould, J. Monograph of the Trochilidæ. Folio. London, 1850.

Gould, J. Handbook of the Birds of Australia. 2 vols., 8vo. London, 1865.

Gray, G. R. List of the specimens of Birds in the British Museum. 8vo. London, 1844-56.

Gray, G. R. The Genera of Birds. 2 edit., 3 vols., fol., 350 pls. London, 1844-49.

Gray and Sharpe. Birds in Voyage Erebus and Terror. London, 1846-1875.

Hartlaub, G. Revision of genus Fulica. Jour. für Ornithologie. i. 1853. (Ger.)

Heuglin, Th. von. Ornithology of North East Africa. 8vo. Cassel, 1869-75. (Ger.)

Hewitson, W. C. Colored illustrations and descriptions of the eggs of British Birds. 2 vols., 8vo. London, 1846. 137 plates.

Horsfield and Moore. Catalogue of Birds in the East India Company's Museum. 2 vols., 8vo. London, 1752-58.

Ingersoll, E. Nests and Eggs of American Birds. Salem, 1879.

Latham, J. Natural History or general Synopsis of Birds. 10 vols., 4to. London, 1781-1802.

Layard and Sharpe. The Birds of South Africa. 1875.

Levaillant, F. Natural History of Paroquets cont'd by St. Hilaire and Souancé. 4 vols., 4to, 300 pls. Paris, 1801-57. (Fr.)

Levaillant, F. Natural History of the Birds of Paradise, Toucans, etc. 2 vols., 114 pls. Paris, 1806. (Fr.)

Maynard, C. J. Birds of Florida. 4to. Salem, 1872-

Merriam, C. H. Review of the Birds of Connecticut. Trans. Conn. Acad., iv. 1877.

Milne Edwards and Grandidier. Natural History of the Birds of Madagascar. 3 vols., 4to. Paris, 1876-78. (Fr.)

Minot, H. D. Land and Game Birds of New England. 8vo. Salem, 1876.

Nuttall, T. Manual of Ornithology of the U. S. and Canada. Boston, 1832-34. 2nd edit., 1840.

Peabody, B. O. Birds of Massachusetts. 8vo. Boston, 1839.

Reichenbach, H. G. L. Natural System of Birds. 4to. Dresden, 1850. (Ger.)

Reichenbach, H. G. L. The Natatores. 4to, 113 pls. Dresden. (Ger.)

Reichenbach, H. G. L. The Grallatores. 4to, 75 pls. Dresden. (Ger.)

Reichenbach, H. G. L. The Fulicariæ et Rallariæ. 4to, 34 pls. Dresden. (Ger.)

Reichenbach, H. G. L. Complete Natural History of the Columbariæ. 4to, 74 pls. Dresden. (Ger.)

Reichenbach, H. G. L. Illustrations of the Gallinaceæ. 4to, 112 pls. Dresden. (Ger.)

Reichenbach, H. G. L. Monograph of the Alcedinæ. 4to, 44 pls. Dresden. (Ger.)

Reichenbach, H. G. L. Monograph of the Sittinæ. 4to, 43 pls. Dresden. (Ger.)

Reichenbach, H. G. L. Monograph of the Tenuirostres. 4to, 62 pls. Dresden. (Ger.)

Reichenbach, H. G. L. Monograph of the Picinæ. 4to, 66 pls. Dresden. (Ger.)

Reichenbach, H. G. L. Monograph of the Trochilinæ. 4to, 176 pls. Leipzig. 1855-62. (Ger.)

Reichenbach, H. G. L. Exotic Song Birds. 4to, 50 pls. Dresden, 1861. (Ger.)

Reichenow. Conspectus Psittacorum. Jour. f. Ornith., xxix. 1881.

Ridgway, R. Outlines of a Natural Arrangement of the Falconidæ. Bull. U. S. Geol. Survey, ii. 1875.

Ridgway, R. Studies of the American Falconidæ. Bull. U. S. Geol. Survey, ii. 1876.

Roux, P. Ornithology of Provence. 2 vols., 4to, 413 pls. Marseilles, 1825-29. (Fr.)

Rowley, G. D. Ornithological Miscellany. 3 vols., 4to, 115 pls. London, 1875-78.

Sagra and d'Orbigny. Mammals and Birds in Sagra's Hist. of Cuba. 4to. Paris, 1840. (Fr.)

Salvin, O. On the Avifauna of the Gallapagos Archipelago. 4to. London, 1876.

Samuels, E. A. Birds of New England. 8vo. Boston, 1870.

Sclater, L. P. Synopsis of the Tanagridæ. Proc. Zoöl. Soc. of London, xxiv. 1856.

Sclater, L. P. Monograph of the Tanagrine Genus Calliste. 8vo, 45 pls. London, 1857.

Sclater and Salvin. Nomenclator avium Neotropicalum. 4to. London, 1873.

Sclater and Salvin. Exotic Ornithology (South American Birds.) 13 parts, 4to, 100 plates. London, 1866-69.

Schlegel and Pollen. Mammals and Birds of Madagascar 4to, 40 pls. Leyden, 1868. (Fr.)

Sharpe, R. B. Monograph of the Alcedinidæ. 4to, 120 pls. London, 1869-71.

Sharpe, R. B. Catalogue of the Accipitres in the British Museum. 8vo. London, 1874.

Sharpe, R. B. Catalogue of the Striges in Brit. Museum. 8vo. London, 1875.

Sharpe, R. B. Catalogue of Passeriformes in British Museum. 8vo. London, 1877.

Stimpson, W. Illustrations of North American Birds. (Trans. Chicago Acad. I, 1867.)

Swainson, W. Birds in Fauna Boreali-Americana. 4to. London, 1831.

Swainson, W. Synopsis of Birds discovered in Mexico. Philosoph. Magazine, 1827.

Swainson, W. Birds of Western Africa. 2 vols., 8vo. London, 1861-62.

Temminck and Laugier. A New Collection of Colored Plates of Birds. 5 vols., 4to. Paris, 1838. (Fr.)

Vieillot, L. P. Natural History of Birds of North America. 2 vols., fol. Paris, 1807. (Fr.)

Vigors, N. A. Birds in Beechey's Voyage. 4to. London, 1839.

Wilson, A. American Ornithology. (1808-1825.) Continued by C. L. Bonaparte. 13 vols., 4to. Phila. 1808-33.

Wilson, A. Jardine's Edition. London and New York, 1876.

REPTILIA AND BATRACHIA.

[It has been found more convenient to catalogue these two divisions together, since in the older works the Batrachia are considered as belonging to the Reptilia.]

Agassiz, L. North American Testudinata. Contributions to the Nat. Hist. of the U. S., vol. ii, 1857.

Allen, J. A. Catalogue of Reptiles and Batrachia of Massachusetts. Springfield, 1879

Baird, S. F. North American Tailed Batrachia.
Jour. Phila. Acad. 4to. 1849.

Baird, S. F. Reptilia in Report Mexican Boundary Survey. 4to. 1849.

Baird, S. F. Reptilia in Reports Pacific Railroad Surveys. Vols. v and x, 4to. 1857-59. (4 papers.)

Baird and Girard. Reptilia in Stansbury's Salt Lake Report. 8 vo. Washington, 1852.

Baird, S. F. Catalogue of North American Serpents. (Smithsonian Misc. Coll., ii. 1853.)

Bell, T. Monograph of the Testudinata. Folio. London, 1836-41.

Bell, T. History of British Reptiles. 2nd edit. 8vo. London, 1849.

Blanford, W. T. Reptilia in Scientific Results second Yarkand Mission. Calcutta, 1878.

Boie, H. Herpetology of Java, in Ferrusac's Bulletin. 1826. (Fr.)

Cocteau et Bibron. Herpetology in de la Sagra's History of Cuba. Paris, 1840.

Cooper, J. G. Reptilia of Washington Territory. Pacific R. R. Survey, vol. xii, pt. ii. 1860.

Cope, E. D. Catalogue of the Venomous Serpents. Proc. Phila. Acad., 1859.

Cope, E. D. Catalogue of the Colubridæ in the Museum of the Phila. Acad. Proc. Acad., 1860.

Cope, E. D. Observations on the Reptiles of the Old World. Proc. Phila. Acad., 1868.

Cope, E. D. On the Arciferous Anura and the Urodela. Jour. Phila. Acad., 1866.

Cope, E. D. On the Families of the Raniform Anura. Jour. Phila. Acad., 1867.

BIBLIOGRAPHY.

Cope, E. D. Synopsis of the Chelydrinæ. Proc. Phila. Acad., 1872.

Cope, E. D. Review of the species of Plethodontidæ and Desmognathidæ. Proc. Phila. Acad., 1869.

Cope, E. D. On the Batrachia and Reptilia of Costa Rico. Jour. Phila. Acad., 1875.

Cope, E. D. Check List of North American Batrachia and Reptilia. (Bull. U. S. Nat. Mus. No. 1, 1875.)

Cope, E. D. On the Primary divisions of the Salamandridæ. Proc. Phila. Acad., 1859. (See also other Papers in Proceedings Journal Phila. Acad., Trans. Am. Philosoph. Soc., and Publications of Hayden's Survey.)

Daudin, F. Natural History of Reptiles. 8 vols., 8vo. Paris, 1802-3. (Fr.)

DeKay, J. E. Natural History of New York. Reptilia. 4to. Albany, 1842.

Dumeril, A. Reptiles and Fishes of West Africa. Paris, 1861. (Fr.)

Dumeril et Bibron. General Herpetology. 10 vols., 8vo, 120 pls. Paris, 1834-54. (Fr.)

Dumeril et Boucourt. Reptiles and Batrachia in Mission Sci. au Mexique et dans l'Amerique Centrale. 4to, Paris, 1870. (Fr.)

Duvernoy, G. L. Natural History of Reptiles. 4to, 46 pls. Paris, 1850. (Fr.)

Girard, C. Monograph of Phrynosoma (Horned Toads). Stansbury's Salt Lake Report. Washington, 1825.

Goode, G. B. Preliminary Catalogue of the Reptiles and Fishes of the Bermudas. Am. Jour. Sci., 1877.

Gray, J. E. Genera of Reptiles. Annals of Philosophy, 1825.

Gray, J. E. Synopsis Reptilium. Pt. 1, Cataphracta (all pub.) 8vo. London, 1831.

Gray, J. E. Reptilia in Beechey's Voyage. 4to. London, 1839.

Gray, J. E. Catalogue of Tortoises, Crocodiles and Amphisbænians in the British Museum. 12mo. London, 1844.

Gray, J. E. Catalogue of Lizards in British Museum. 12mo. London, 1845.

Gray, J. E. Catalogue of Snakes in British Museum. London, 1849.

Gray, J. E. Catalogue of Batrachia Gradentia in British Museum. 12mo. London, 1850.

Gray, J. E. Descriptive Catalogue of Tortoises in the British Museum. 2 vols. and Suppl. 4to, 50 pls. London, 1855-70.

Gray, J. E. Synopsis of the recent Crocodilians. 4to. London, 1867.

Gray, J. E. Hand list of Shield Reptiles in British Museum. 8vo. London, 1873.

Gray and Sowerby. Monograph of the Testudinata. 4to, 60 pls. London, 1872.

Guichenot. Reptilia in l'Exploration scientifique de l'Algerie. 4to. Paris, 1850. (Fr.)

Guichenot. New or Rare Reptiles from South America (in Castelnau's Expedition). 4to. Paris, 1855.

Gunther, A. Catalogue of the Batrachia Salentia in the British Museum. 8vo. London, 1858.

Gunther, A. Catalogue of the Colubrine Snakes in the British Museum. 12mo. London, 1858.

Gunther, A. Reptiles of British India. (Ray Society.) fol., 26 pls. London, 1864.

Hallowell, E. Reptiles of the Zuni and Colorado Rivers (in Sitgreaves' Report). Washington, 1853.

Hallowell, E. New Reptiles from Oregon and West Africa. (Jour. Phila. Acad.) 1854. 4to.

BIBLIOGRAPHY. 167

Hallowell, E. Reptilia in Reports Pacific R. R. Survey. Vol. x. 1859.

Hoffman, C. K. The classes and orders of the Amphibia. [Bronn's "Klassen und Ordnung."] 1 vol., 8vo. Leipzig, 1874–77.

Holbrook, J. E. North American Herpetology. 5 vols., 4to. Phila., 1836–42.

Jan, G. Iconography of the Ophidia, continued by Sordelli. 50 parts, 4to., 300 plates. Paris, 1860–76. (Fr.)

Leconte, John. Catalogue of North American Testudinata. Proc. Phila. Acad., 1854.

Peters, W. On the classification of the Cæcilians (Monatsber. Berlin Acad., 1879). 8vo. (Ger.)

Schinz, H. R. Natural History and Illustrations of Reptiles. fol., 102 pls. Leipzig, 1833. (Ger.)

Schlegel, A. Illustrations of New or little-known Amphibia. Text. 8vo, atlas, 50 fo'. plates. Düsseldorf, 1837–44. (Ger.)

Spix, J. B. New Species of Tortoises and Frogs collected in Brazil. 4to, 39 pls. Munich, 1824. (Lat.)

Steindacher, Franz. Amphibia and Reptilia of the "Novara" Expedition around the world. 4to. Vienna, 1867. (Ger.)

Storer, D. H. Reptiles of Massachusetts. 8vo. Boston, 1839.

Stranch, A. Revision of the Genera of Salamanders with desc. of new sp. 4to. St. Petersburg, 1873.

Swainson, W. Natural History of Fishes, Reptiles and Amphibians. 2 vols., 8vo. London, 1838–39.

Tschudi, J. J. Investigations of the Peruvian Fauna (Mammals and Reptiles). 4to, 30 pls. St. Gall, 1844–46. (Ger.)

Wiegmann, A. F. Herpetologia Mexicana. Berlin, 1834.

Yarrow, H. C. Reptiles and Batrachi in Wheeler's Survey. 4to, vol. v. Washington, 1875.

Fishes.

Agassiz, L. Natural History of the fresh-water fishes of central Europe. 2 pts., 8vo, with folio atlas, 52 plates. Neuchatel, 1839-42. (Fr.)

Agassiz, L. Selected genera and species of Fish collected by Spix in Brazil. fol., 96 pls. Munich, 1829. (Lat.)

Ayers, W. O. Fishes from Brookhaven, L. I. (Jour. Boston Soc., 1 vol.). 1842.

Baird, S. F. Fishes of New Jersey and Long Island. (9th Report Smithsonian Inst.) • 1855.

Baird and Girard. Fishes in Sitgreaves' Zuni and Colorado Expedition Report. 8vo. Washington, 1853.

Beavan, R. Fresh-water Fishes of India. 8vo. London, 1877.

Bennett, J. W. Fishes of Ceylon. 4to. London. 1828-30.

Bleeker, P. Prodrome of the Ichthyology of the Indian Archipelago. 4to. Batavia, 1858-60. (Lat.)

Bleeker, P. Ichthyological Atlas of the Dutch East Indies (33 parts published). Fol., 500 pls. Amsterdam, 1862-77. (Fr.)

Bleeker et Pollen. Fishes of Madagascar and Reunion. 4to. Leyden, 1874.

Bloch, M. E. Natural History of Native and Exotic Fishes. 4to, 432 pls. Berlin, 1782-95. (Ger.)

Bloch, M. E. Ichthyology, a general and specific history of Fish. 12 vols., fol., 432 pls. Berlin, 1785-97. (Fr.)

Bonnaterre. Natural History of Fish. 4to, 100 pls. Paris, 1788. (Fr.)

Brevoort, J. C. Notes on Japanese Fish. (Perry's Japan Expedition Report. Vol. ii.) Washington, 1857.

Canestrini, J. Systematic arrangement of the Percidæ (Verhandlung Zoöl.-Bot. Gesellschaft in Wien). 1860. (Ger.)

Cantor, T. Catalogue of Malayan Fishes. 8vo. Calcutta, 1850.

Castelnau, F. New or rare fishes from South America. 4to, 50 pls. Paris, 1855. (Fr.)

Cope, E. D. Synopsis of the Cyprinidæ of Pennsylvania. (Trans. Am. Phila. Soc., xiii.) 4to. Phila., 1866.

Cope, E. D. Partial synopsis of the Fresh water fishes of North Carolina. 8vo. Proc. Am. Phil. Soc., 1870. 2nd edit., 1877.

Cope, E. D. Contributions to the Ichthyology of the lesser Antilles. (Trans. Am. Phila. Soc., 1871). 4to.

Cope, E. D. Report on Reptiles and Fishes. Rep. U. S. Geol. Survey for 1870.

Cope, E. D. On the Fishes of the Ambyiacu River. (Proc. Phila. Acad., 1871.)

Cope, E. D. Contribution to the Ichthyology of Alaska (Proc. Am. Phila. Soc., 1873).

Cope and Yarrow. Fishes, in Zoölogy of Wheeler's Survey. vol. v, 4to. Washington, 1875.

Cuvier et Valenciennes. Natural History of Fishes. 22 vols., 4to, with atlas, 650 plates. Paris, 1829-49. (Fr.)

Cuvier et Valenciennes. The same, 8vo.

Day, F. Fishes of Malabar and Cochin China. 4to. London, 1865.

Day, F. Fishes of India. 4 pts., 4to, 160 pls. London, 1875-8.

DeKay, J. E. N. Y. Fauna: Fishes. 4to, 79 pls. Albany, 1842.

Dumeril, A. Reptiles and Fishes of West Africa. 4to. Paris, 1861. (Fr.)

Dumeril, A. Ichthyology or Natural History of Fishes. 3 vols., 8vo. Paris, 1850-7. (Fr.)

Fitzinger, L. J. Atlas of the Natural History of Fish. 4to, 77 pls. Vienna, 1864. (Ger.)

Gill, Theodore. Catalogue of Fishes of the East Coast of N. A. (Rep. U. S. Fish Com. for 1871-2). Washington, 1875. (See also numerous other papers, Proc. Phila. Acad., 1863 et seq.)

Girard, Charles. Monograph of the Fresh-water Cottoids of North America. (Smithsonian Contributions iii) 4to. 1851.

Girard, Charles. Fishes in Reports Pacific R. R. Survey. Vols. vi and x. 4to. Washington, 1857-59.

Girard, Charles. Ichthyology of the Mexican Boundary Survey. 4to. Washington, 1859.

Goode, G. Brown. Preliminary Catalogue of the Reptiles and Fishes of the Bermudas. (Am. Jour. Sci., 1877.)

Goode and Bean. List of the Fishes of Essex Co., Mass. (Bull. Essex Inst. xi, 1879).

Goode and Bean. Notices of 50 species East Coast Fishes. (Am. Jour. Sci. xvii, 1879.)

Gray, J. E. List of Cartilaginous Fishes in the British Museum. 8vo. London, 1851.

Gray and Gerrard. List of Chondropterygii in the British Museum. 12mo. London, 1851.

Guichenot. Fishes in de la Sagra's Hist. Cuba. Paris, 1854. (Fr.)

Gunther, A. Catalogue of the Fishes in the British Museum. 8 vols., 8vo. London, 1859-70. (Gives diagnoses of all known forms. Indispensable.)

Gunther, A. Fishes of Central America. 4to, 25 pls. London, 1868.

Gunther, A. Fishes of the South Seas. 4to, many plates. Hamburg, 1873. (Ger.)

Hamilton, F. Fishes of the Ganges. 4to, atlas fol. Edinburg, 1822.

Holbrook, J. E. Southern Ichthyology. No. ii. 4to. New York, 1847 (no more published).

Holbrook, J. E. Ichthyology of South Carolina, 4to. Charleston, 1855 (incomplete).

Holbrook, J. E. Ichthyology of South Carolina. 4to. Charleston, 1860 (incomplete). The text is essentially the same but the plates are different.

Holmes, E. Report on the Fishes of Maine [etc.]. 2nd Rep. on the Geol. and Nat. Hist. of Maine, 1862.

Hutton and Hector. Fishes of New Zealand. 8vo. Wellington, N. Z., 1872.

Jenyns, L. Fishes in Voyage of the Beagle. 4to. London, 1842.

Jordan, D. S. Synopsis of the genera and species of fish to be looked for in Indiana. (Rep. Ind. Geol. Sur., 1874 [75].)

Jordan, D. S. Catalogue of the Fishes of Illinois. (Bull. Ill. Industrial Univ., No. 2, 1878.)

Jordan, D. S. Fishes of Northern Indiana. (Proc. Phila. Acad., 1877.)

Jordan, D. S. Partial Synopsis of the Fishes of upper Georgia [etc.]. Annals N. Y. Lyceum, xi, 1877.

Jordan, D. S. Contributions to North American Ichthyology, No. 1. Bulletin U. S. Nat. Mus. 9, 1877.

Jordan, D. S. On a collection of Fishes made in Dacota, and Montana. (Bulletin U. S. Geol. Sur. iv, 1878.)

Jordan, D. S. Concerning Fishes of "Ichthyologia ohioensis." (Bull. Buff Soc , 1876.)

Jordan, D. S. Synopsis of the Catostomidæ. (Bull. Nat. Mus. 12, 1878.)

Jordan, D. S. Catalogue of the Fishes of the Fresh waters of North America. Bull. U. S. Geol. Survey iv, 1878.

Jordan and Copeland. Check list of the fishes of the fresh waters of North America. (Bull. Buff. Soc., 1877.)

Jordan and Gilbert. On the genera of North American Fresh-water Fishes. (Proc. Phila. Acad., 1877.)

Kaup, J. J. Catalogues of the Apodal and Lophobranchiate fish in the British Museum. 2 vols., 8vo. London, 1856.

Knight, T. F. Descriptive Catalogue of fishes of Nova Scotia, Halifax. 8vo. 1866.

Kroyer, H. Danish Fish. 4 pts., 8vo. Copenhagen, 1838–53. (Danish.)

Kroyer, H. Fishes in Gaimard's Travels in Scandinavia. Folio. Paris, 1847.

Lay and Bennett. Fishes of Beechey's Voyage. London, 1839.

Lesueur, C. A. Papers in Jour. Phila. Acad., 1817–1822.

Love. Fishes of Madeira. 5 pts., 8vo. London, 1843–60.

McClellan, J. Indian Cyprinidæ. 4to. Calcutta, 1839.

Mitchell, S. L. The Fishes of New York. (Trans. Lit. and Philos. Soc., N. Y. i, 1815.)

Mitchell, S. L. Memoir on Ichthyology. (American Monthly Mag. and Critical Review, ii, 1818.)

Müller and Henle. Systematic descriptions of the Plagiostomi. fol. Berlin, 1841. (Ger.)

Müller and Troschel. Horæ Ichthyologiæ. 3 pts., fol. Berlin, 1845-49. (Ger.)

Playfair and Günther. Fishes of Zanzibar. 4to. London, 1877.

Poey, F. Conspectus of Cuban Fish. (Mem. Hist. Nat. Cuba, ii.) 8vo. Havana, 1861.

Poey, F. Synopsis of Cuban Fish. (Repertorio fisico-natural de la isla Cuba, ii.) 8vo. Havana, 1868.

Poey, F. Review of Cuban Trisotropes. (Ann. N. Y. Lyceum, ix, 1870.)

Poey, F. Monograph of the Cuban Sparini. (Ann. N. Y. Lyceum, x, 1872.) (Fr.)

Putnam, F. W. Fishes of Essex Co., Mass. (Proc. Essex Inst., i, 1855-6.)

Putnam, F. W. List of Fishes sent by the Museum to other Institutions. (Bulletin Mus. Comp. Zoöl., i, 1863.)

Richardson. Fishes in Fauna Boreali Americana. 4to. London, 1836.

Schinz, H. R. Natural History and Illustrations of Fish. folio, 97 pls. Leipzig, 1836. (Ger.)

Schomburgh, R. H. Natural History of the Fishes of Guiana. 2 vols., 8vo, 60 pls. Edinburg, 1841-43.

Siebold, Temminck and Schlegel. Fishes of the Fauna Japonica. folio. 161 plates. Leyden, 1844-50.

Steindacher, F. Many papers in Verhandlung Zoöl. bot. Gesellschaft Wein, and other Austrian publications.

Storer, D. H. Report on the Fishes of Massachusetts. 8vo. Boston, 1839 (almost the same under the same title in Jour. Bost. Soc. ii-iv, 1839-1842).

Storer, D. H. Synopsis of the Fishes of North America. 4to. Cambridge, 1846. (Ext. from Memoirs Am. Acad. Arts and Sci. New Series, ii, 1846.

Storer, D. H. A History of the Fishes of Massachusetts. (Memoirs Am. Acad. New Series, vols. v-ix. 39 pls. 4to. 1853-1867.)

Suckley, G. [Fishes of Washington Territory]. (Pacific R. R. Survey). Vol. xii, pt. ii, 1860.

Vaillant, L. The Etheostomoids of North American Fresh waters. (Nouv. Archives du Mus.) 4to. Paris. 1873. (Fr.)

Vaillant et Boucourt. Fishes of the Mission Scientifique au Mexique et dans l'Amerique Centrale. 4to. Paris, 1874. (Fr.)

Valenciennes, A. Ichthyology of the Canary Isles. 4to. Paris, 1842.

Valenciennes, A. Fishes in D'Orbigny's Voyage to Central America. 4to. Paris, 1847.

Wright, Fries, and Ekström. Scandinavian Fish. 4to. 63 pls. Stockholm, 1836-57. (Swedish.)

Yarrell, W. History of British Fishes. 2 vols. and Suppl. 8vo. London, 1836–39. 3d edit. 2 vols. Lond. 1859–60.

INVERTEBRATA (General).

Guerin-Meneville. Articulata in Sagra's Cuba. Text, 8vo. atlas, folio. Paris, 1857. (French).

Gould, A. A. Report on the Invertebrata of Massachusetts. 8vo. Cambridge, 1841.

Lamarck, J. B. Natural History of Invertebrated Animals. 7 vols. 8vo. Paris, 1815–1822. (Fr.)

Lamarck, J. B. Second edition by Deshayes and Milne Edwards. 11 vols. 8vo. Paris, 1835–45. (Fr.)

Latreille, P. A. Natural History of Crustacea and Insects. 14 vols. 8vo. Paris, 1802–05. (Fr.)

Latreille, P. A. Genera of Crustacea and Insects. 4 vols. 8vo. Paris, 1806–9. (Fr.)

Leidy, Joseph. Marine Invertebrate Fauna of Rhode Island and New Jersey. (Jour. Phila. Acad. 1855.)

Leuckart, R. Human Parasites. 2 vols. 8vo. Leipzig, 1863–76. (Ger.)

Lucas, H. Natural History of Crustacea, Arachnida and Myriapoda. 8vo. Paris, 1840. (Fr.)

Lucas, H. Articulata in Expl. Scientifique de l'Algerie. 4 vols. 4to. Paris, 1846–50. (Fr.)

Packard, A. S., jr. List of Animals dredged in Southern Labrador. (Canadian Naturalist and Geologist. iii, 1863.)

Packard, A. S., jr. Recent Invertebrate Fauna of Labrador (Memoirs Bost. Soc. 4to. i. 1867.)

Sars, G. O. Remarkable forms from great depths off the Norwegian Coast. 4to. 13 pls. Christiana, 1872–75.

Sars, M. and G. O. Contributions to a knowledge of the **Fauna of Christiana Fjord. 3 pts. 8vo. Christiania, 1868–73.**

Sars, Koren and Danielssen. Fauna of the shores of Norway. 3 vols., 38 pls., 4to. Bergen, 1846–1877.

Smith and Harger. Report on the Dredgings in the region of St. George's Banks, in 1872. (Trans. Conn. Acad. iii. 1874.)

Stimpson, W. Marine Invertebrata of Grand Menan. (Smithsonian Contributions. vi. 1857.)

Verrill, A. E. External and Internal Parasites of man and domestic animals. (Rep. Conn. Board of Agriculture, 1870.)

Verrill, A. E. New Invertebrata of the N. E. Coast of the U. S. (Proc. U. S. National Museum, i, 1879.) (See also numerous papers in Am. Jour. Sci.)

Verrill, A. E. Preliminary Check-list of Invertebrata of North East Coast of America. 8vo. New Haven, 1879.

Verrill, Smith and Harger. Marine Invertebrata of Vineyard Sound. (Rep. U. S. Fish Commission. 1871-2 [1873].)

INSECTA (General).

Audouin and Brulle. Natural History of Insects. 4 vols. 8vo. Paris, 1834–37. (Fr.)

Beauvois, de, A. M. and F. J. P. Insects from Africa and America (etc.). 4to. 90 pls. Paris, 1805. (Fr.)

Bedel et Simon. Cave Insects. 8vo. Paris, 1875. (Fr.)

Blanchard. Natural History of Insects. 3 vols. 8vo. Paris, 1840–41. (Fr.)

Blanchard. Metamorphoses, Habits and Instincts of Insects. 1 vol. 8vo. Paris, 1866. (Fr.)

Blanchard. Same translated by Duncan. London, 1868.

Boisduval, J. A. Entomological Fauna of Oceanica. 8vo. Paris, 1835. (Fr.)

Burmeister, H. Handbook of Entomology. 5 parts. 8vo. Berlin, 1832-55. (Ger.)

Burmeister, H. Same (Manual of Entomology), translated by Shuckard. 1 vol. 8vo London, 1836.

Burmeister, H. Illustration of the Genera of Insects. 8vo. Berlin, 1838-46.

Catlow, M. E. Popular British Entomology. 1 vol. 12 mo. London, 1860.

Curtis, J. British Entomology. 3 vols. London.

Donovan. Natural History of the Insects of China. 4to. London, 1798.

Donovan. New Edition by Westwood. London, 1842.

Donovan. Natural History of the Insects of India. (Westwood's Edition.) 4to. London, 1838.

Drury, Dru. Illustrations of Exotic Entomology. 3 vols. 4to. London, 1770-1782.

Drury, Dru. New Edition by Westwood. London, 1837.

Emmons and Fitch. New York Fauna. Vol. v. Insects. 4to. Albany, 1854.

Fabricius, J. C. Species Insectorum. 2 vols. 8vo. Hamburg, 1781. (Lat.)

Fabricius, J. C. Mantissa Insectorum. 2 vols. 8vo. Copenhagen, 1787. (Lat.)

Fabricius, J. C. Entomologiæ Systematicæ with Suppl. and Index. 7 vols. 8vo. Copenhagen, 1792-98.

Fitch, A. Reports on the Noxious and Beneficial Insects of the State of New York. 14 nos. (Trans. N. Y. State Agric. Soc. 1854-68.)

Geer, de, C. Memoirs on the History of Insects. 7 vols. 4to. Stockholm, 1752-78. (French.)

Germar et Ahrens. European Insect Fauna. 6 vols. 12mo. 575 plates. Halle, 1813-1818. (Lat.)

Glover, T. Notes from my Journal. 4to. Washington (privately printed).

Guerin et Percheron. Genera of Insects. 8vo. Paris, 1835-38. (Fr.)

Hagen, H. A. Biblotheca Entomologica. (Bibliography of Entomology. 2 vols. 8vo. Leipzig, 1862-63.

Haldemann. Insects in Stansbury's Salt Lake Report. 8vo. Washington, 1852.

Harris, M. The Aurelian. New Edition by Westwood. 4to. London, 1840.

Harris, T. W. Insects injurious to Vegetation. 8vo Cambridge, 1842.

Harris, T. W. New edition by Flint. Boston, 1862.

Harris, T. W. Entomological Correspondence, edited by Scudder. 8vo. Boston, 1869.

Kirby, W. Insects of the Fauna boreali Americana. 4to. London, 1836. (Reprint by Bethune in Canadian Entomologist, ii, 1870 *et seq.*)

Kirby and Spence. Introduction to Entomology. 2d edit., 2 vols. 8vo. London, 1818. 5th edit. 4 vols., 1828. 7th (comprising vols. 3 and 4 of 5 edit.,) 1 vol. 8vo. London, 1857.

Lacordaire, T. Introduction to Entomology. (Suites à Buffon.) 2 vols. 8vo. Paris, 1834-38. (Fr.)

Lintner, J. A. Papers in Reports Regents. N. Y. University.

Lucas, H. Entomology of Castelnau's So. American Voyage. 4to. Paris, 1857. (Fr.)

MacLeay, W. S. Annulosa Javanica. 4to. London, 1825. (French Edition.) Paris, 1833.

MacLeay, W. Illustrations of the Annulosa of South Africa. 4to. London, 1838.

Müller, J. Terminologica Entomologica. (Dictionary of Entomological terms.) 2 edit. 12mo. Brünn, 1782.

Packard, A. S. Record of American Entomology. 5 parts. 8vo. Salem, 1869-73.

Packard, A. S. Guide to the Study of Insects. 1st edit. 8vo. Salem, 1869. 8th edit. N. Y., 1880.

Packard, A. S. Report on Rocky Mt. Locust and other Insects. (Rep. U. S. Geol. Surv., 1875, [1877]).

Packard, A. S. Reports on Injurious and Beneficial Insects of Mass. (Mass. Agr. Repts., 1871-73.)

Packard, A. S. Directions for Collecting and preserving Insects. (Smithsonian Misc. Coll. xi.)

Panzer, G. W. German Insect Fauna. Continued by Herrich-Schäffer and Roch. 190 parts, 12mo., 4572 plates. Nuremburg and Regensburg, 1793-1844.

Ratzeburg, J. T. Forest Insects. 3 vols., 4to. Berlin, 1839-44. (Ger.)

Riley, C. V. Missouri State Entomological Reports. 1869 et seq.

Savigny, J. C. Entomological Fauna of Egypt. fol. Paris, 1818. (Fr.)

Say, T. American Entomology. 3 vols. 8vo. Phila., 1824-28.

Say, T. Complete writings on Entomology. Edited by Leconte. 2 vols., 8vo. N Y. 1859.

Spix et Martius. Articulata collected in Brazil. Folio. Munich, 1830-34. (Lat.)

Stephens, J. F. Illustrations of British Entomology. 12 vols , 100 pls., 8vo. London, 1828-46.

Walker, F. Insecta Saundersiana. 8vo. London, 1856.

Walsh, Lebaron and Thomas. Reports Ill. State Entomologists.

Westwood, J. O. Introduction to the Modern Classification of Insects. 2 vols., 8vo. London, 1839-40.

Westwood, J. O. Arcana Entomologica. 2 vols., 8vo, 96 pls. London, 1845.

Westwood, J. O. Thesaurus Entomologica Hopeianus. Fol. Oxford, 1875.

White and Butler. Insects of the voyage of the Erebus and Terror. 4to. London, 1874.

Wollaston, T. Insecta Madeirensia. 4to. London, 1874.

HYMENOPTERA.

Blake, C. J. Synopsis of Mutillidæ of North America. (Trans. Am. Ent. Soc., 1871.)

Buckley, S. B. Descriptions of New Species N. A. Formicidæ. Phila., 1866-7.

Cresson, E. T. Papers in Proc. Phila. Acad., and Trans. Amer. Ent. Soc., Canadian Entom., Proc. Bos. Soc. 1863 et seq.

Cresson. Hymenoptera texana. Trans. Am. Ent. Soc., iv, 1872.

Cresson and Norton. Hymenoptera in Report Wheeler's Survey. Vol. v. 1875.

Giraud, J. Papers in Verhandl. Zoöl. Bot. Gesellschaft in Wien, 1858 et seq. (Ger.)

Gravenhorst, J. L. C. Icneumonologiæ Europæa. 3 vols., 8vo Ratisbonne, 1829.

Green, J. W. Review of the American Bombidæ. (Ann. N. Y. Lyceum, vii, 1860.)

Haliday, A. H. Hymenoptera Brittanica. 8vo. London, 1839.

Hartig, T. Families of Leaf and Wood Wasps. 8vo. Berlin, 1837. (Ger.)

Kirby, W. Monographia Apium Angliæ (Monograph of British Bees. 2 vols., 8vo. Ipswich, 1802.

Kirchner, L. Catalogue of the Hymenoptera of Europe. 8vo. Vienna, 1867.

Latreille, P A. Bees of Central America. 4to. Paris, 1811. (Fr.)

Lepelletier, St. Fargeau et Brulle. Natural History of the Hymenoptera. (Suites à Buffon). 4 vols., 8vo. Paris, 1836-45. (Fr.)

Linden, P. L. van der. Fossorial Hymenoptera of Europe. 4to. Brussels, 1829. (Fr.)

Marshall, T. A. Catalogues of British Hymenoptera. 8vo. London, 1873.

Mayr, G. L. Synonymical Index of Formicidæ. (Verh. Zoöl. Bot. Gesell. in Wien. 1863)

Mayr, G. L. Australian Formicidæ 4to. Hamburg, 1876. (Ger.) (See other papers on Formicidæ, etc., Vienna publications, 1850.)

Moggridge, J. T. Harvesting ants and Trap-door spiders. 2 vols., 8vo. London, 1873-74.

Norton, E. Catalogue of Tenthredinidæ and Uroceridæ of N. A. (Trans. Am. Ent. Soc., 1, 1867-8.)

Norton, E. Description of Mexican Ants. (Proc. Essex Inst., vi. 1868.

Norton, E. Hymenoptera of the Genus Atlantus in the U. S. (Jour. Bost. Soc., vii. 1860.)

Ormerod, E. L. British social wasps. 8vo. London, 1868.

Osten-Sacken. Cynipidæ of the N. A oaks and their galls. (Phila., 1861 5,) and (Trans. Am. Ent. Soc., iii, 1870).

Packard, A. S., jr. Revision of the Fossorial Hymenoptera of N. A. (Proc. Ent. Soc. Phila., 1865-7.)

Packard, A. S., Jr. Humble Bees of New England and their Parasites. (Proc. Essex Inst., iv, 1864.)

Saussure, H. de. Studies of the Family Vespidæ. 3 vols., 8vo. 75 pls. Geneva, 1852-58.

Saussure, H. de. Synopsis of American Solitary Wasps. (Smithsonian Misc. Coll.) Washington, 1875.

Shuckard, W E. British fossorial Hymenoptera. 8vo. London, 1837.

Shuckard, W. E. Natural History of British Bees. 8vo. London, 1866.

Sichel, J. et Radoszkovsky. Monograph of the Mutillas. (St. Petersburg.) 8vo. 1869-70. (Fr.)

Smith, F. Catalogue of Hymenoptera in British Museum. 7 parts, 12 mo. London, 1853-59. (See also other papers in Trans. Ent. Soc. London, 1853-59. See also other papers in Trans. Ent. Soc., London, etc.)

Thompsen, C. G. Scandinavian Hymenoptera. 4 vols., 8vo Lund, 1871-76. (Lat.)

Walker, F. Monograph of Chalcidæ. 8vo. London, 1839.

Walker, F. List of Chalcidæ in the British Museum. 8vo. London, 1845-8.

Walker, F. Hymenoptera collected in Egypt and Arabia. 8vo. London, 1870.

Walker, F. Notes on Chalcidæ. 7 parts, 8vo. London, 1871-72.

Walsh, B. D. Papers Trans. St. Louis Acad., 1873, and American Entomologist, ii, 1870.

Wesmael, C. Monograph of the Braconidæ of Belgium. 4to. Brussels, 1835-37. (Fr.)

LEPIDOPTERA.

Blanchard and Doyere. Lepidoptera in Regne Animal. 4to 31 pls. Paris, 1849. (Fr.)

Boisduval, J. A. Monograph of Zygænidæ. 8vo Paris, 1829. (Fr.)

Boisduval, J. A. New and little-known Lepidoptera of Europe. 8vo, 81 pls. Paris, 1832-41. (Fr.)

Boisduval, J. A. Species generale des Lepidopteres. 8vo (2 vols. pub.) 34 pls. Paris, 1836-74. (Fr.)

Boisduval et Guenée. Species generale des Lepidopteres. 8 vols., 8vo. (Suites à Buffon.) Paris, 1836-74. (Fr.)

Boisduval et Leconte. Iconography of North American Butterflies. 8vo, 78 pls. Paris, 1830-42. (Fr.)

Butler, A. G. Catalogue of Satyridæ. 8vo. London, 1869.

Butler, A. G. Lepidoptera exotica. 4to, 64 pls. London, 1869-74.

Butler, A. G. Monograph of Callidryas. 4to, 16 pls. London, 1874.

Butler, A. G. Revision of Sphingidæ. 4to. London, 1875.

Butler, A. G. Illustrations of Typical Heterocera. 4to., many plates. London, 1877.

Butler, A. G. Synonymic List of Species of Pieris. (Proc. Zoöl. Soc., 1872.)

Chambers, V. T. Index of described Tineina of North America. (Bulletin Hayden's Survey, 1877.

Clemens, B. Tineina of North America. Edited by Stainton. 8vo. London, 1872. (A collection of all his papers on the subject.)

Cramer, P. Papillons exotiques. 4 vols., 4to, 442 plates. Amsterdam, 1789-91. (Fr.)

Doubleday and Gray. List of Lepidoptera in British Museum. 3 vols., 8vo. London, 1844-56.

Doubleday, Hewiston and Westwood. Genera of Diurnal Lepidoptera. 2 vols., fol., many plates. London, 1846-52.

Duncan and Westwood. History of British Lepidoptera. 2 vols., 8vo. Edinburgh, 1836.

Duncan and Westwood. Exotic Lepidoptera. 2 vols., 8vo. 67 pls. Edinburg, 1842.

Duponchel, P. A. Catalogue Lepidoptera of Europe. 8vo. Paris, 1844.

Duponchel and Guenée. Iconography of the Caterpillars. 2 vols., 8vo, 92 pls. Paris, 1849. (Fr.)

Edwards, H. Papers in Proc. Cal. Acad.

Edwards, W. H. Butterflies of North America. 4to, many pls. Phila. 1868.

Edwards, W. H. New species North American Butterflies. (Trans. Am. Ent. Soc., iii-iv, 1872.)

Edwards, W. H. Catalogue of Lepidoptera of North America. 8vo Phila., 1877.

Esper, E. J. European Butterflies. Continued by Charpentier. 5 vols., 4to, 439 pls. Erlangen, 1777-1832. (Ger.)

Felder, C. and J. Lepidoptera of Novara's Voyage. 4to, 140 pls. Vienna, 1865-77. (Ger.)

Freyer, C. F. New Contributions to a knowledge of the Lepidoptera. 7 vols., 4to, 700 pls. Augsburg, 1831-59. (Ger.)

Gerhard, B. Monograph European Lycænidæ. 4to, 39 pls. Hamburg, 1853.

Gerhard, B. Systematic Arrangement of North American Macro-Lepidoptera. 8vo. Berlin, 1878.

Gray, G. R. Catalogue Lepidopterous Insects in British Museum. 4to. London, 1852.

Gray, Walker, Stainton, etc. List of Lepidoptera in British Museum. 35 pts., 8vo. London, 1854-65.

Grote, A. R. See numerous short papers in Trans. Am. Ent. Soc., Bulletin Buff. Soc., Bulletin Hayden's Survey, Canadian Entomologist, etc. 1863 *et seq.*

Grote and Robinson. As above.

Harris, T. W. North American Sphingidæ. (Am. Jour. Sci., xxxvi.)

Harvey, L. F. On Texan Lepidoptera. Bull. Buff. Soc., iii, 1877.

Herrich-Schäffer, G. Systematic treatise on European Butterflies. 6 vols., 4to, 670 pls. Regensburg, 1843-56. (Ger.)

Herrich-Schäffer, G. Lepidoptera exotica nova. Regensburg, 1850-58.

Hewlston. Illustrations of Exotic Butterflies. 5 vols., 4to, 300 pls. London, 1861-76.

Hewlston. Illustrations of Diurnal Lepidoptera-Lycænidæ. 4to, 79 pls. London, 1863-75.

Hübner, J. Collection of European Butterflies. Continued by Geyer. 4to, 790 pls. Augsburg, 1805-41. (Ger.)

Hübner, J. History of European Butterflies. 4to, 485 pls. Augsburg, 1806-41. (Ger.)

Humphreys, H. N. Genera of British Moths. 2 vols., 4to, 60 pls. London, 1861.

Kirby, W. F. Manual of European Butterflies. 8vo. London, 1862.

Kirby, F. Synonymic Catalogue of Diurnal Lepidoptera (of the world) with suppl. 8vo. London, 1871-77.

Mead and Edwards. Lepidoptera in Rep. Wheeler's Survey, v, 1875.

Minot, C S. American Lepidoptera, Proc. Bos. Soc., xli, 1869.

Morris, J. G. A Natural History of British Moths. 4 vols., 8vo. London, 1871.

Morris, J. G. Synopsis of Lepidoptera of North America. Pt. 1 (all published). Smithsonian Misc. Coll., iv, 1862.

Newman, E. Illustrated Nat. Hist. British Moths. 8vo. 740 fig. London, 1869.

Packard, A. S., jr. Notes on Zygænidæ. (Proc. Essex Inst. iv, 1864.)

Packard, A. S., jr. Synopsis Bombycidæ of the United States. (Proc. Ent. Soc., Phila., 1864.)

Packard, A. S., jr. Catalogue of Phalænidæ of California. (Proc. Bost. Soc., 1870-74).

Packard, A. S., jr. New American Moths; Zygænidæ and Bombycidæ. (Rep. Peabody Acad., 1872.)

Packard, A. S., jr. Catalogue of Pyralidæ of California. (Ann. N. Y. Lyceum, 1873.)

Packard, A. S., jr. Monograph of the Phalænidæ of the U. S. (Hayden's Survey.) 4to, 13 pls. Washington, 1876.

Robinson, C. T. Notes on American Tortricidæ (Trans. Am. Ent. Soc., ii, 1869.)

Saunders, W. Synopsis of Canadian Arctiadæ. Toronto, 1863.

Scudder, S. H. List of Butterflies of New England. (Proc. Essex Inst., iii, 1863).

Scudder, S. H. Revision of North American species of Chionobas. (Proc. Ent. Soc., Phila., 1865.)

Scudder, S. H. Species of Pamphila. (Memoirs Bost. Soc., ii, 1874.)

Scudder, S. H. Synonymic List of North American Butterflies. (Bulletin Buffalo Soc., 1875-76.)

Scudder, S. H. Systematic Revision of some American Butterflies. (4th Rep. Peabody Acad., 1872.)

Smith and Abbot. Natural History of Rarer Lepidopterous Insects of Georgia. 2 vols., fol., 104 pls. London, 1797.

Stainton, H. T. Insecta Brittanica; Tineina. 8vo. London, 1854.

Stainton, Zeller, Douglas and Frey. Natural History of the Tineina. 13 vols., 8vo, 102 pls. London, 1858-73. (Eng., Lat., Fr., Ger.

Staudinger et Wocke. Catalogue of European Lepidoptera 8vo. Dresden, 1871.

Strecker, H. Lepidoptera, indigenous and exotic. 4to, many pls. Reading, Pa., 1872.

Stretch. Illustrations of Zygænidæ and Bombycidæ. 8vo, pls. San Francisco, 1874.

Westwood. Synopsis of the Uranidæ. Trans. Zoöl. Soc., **x, 1879.**

Wilkinson, S. J. British Tortrices. 8vo. London, 1859.
Zeller, P. C. Contributions to a knowledge of North American Heterocera [Texan forms]. (Verh. Zoöl. Bot. Ges. Wien, 1872-74. Ger.) See numerous other papers in Isis, and publications of Berlin and Vienna Societies, 1838.

DIPTERA.

Bergenstamm und Loew. Synopsis Cecidomyidarum. 8vo. Vienna, 1876.
Bigot, J. Classification of the Diptera. 8vo. Paris, 1852-4. (Fr.)
Bigot, J. Diptera of Madagascar and Gaboon. 8vo. Paris, 1858-9. (Fr.)
Brauer, F. Monograph of the Œstridæ. 8vo. Vienna, 1863. (Ger.)
Fallen, C. F. Diptera Sueciæ. 2 vols., 4to. London, 1814-25.
Loew, H. New contributions to a knowledge of the Diptera. 4to. Berlin, 1853-62. (Ger.)
Loew, H. South African Diptera. 4to. Berlin, 1861. (Ger.)
Loew, H. European Trypetidæ. Folio. Vienna, 1862.
Loew, H. Monographs of the Diptera of North America, edited by Osten Sacken. Parts i, iv. (Smithsonian Misc. Coll., vols., vi, viii, and ix) Washington, 1862-73.
Loew, H. Diptera Americanæ septentrionalis. (Berlin, Ent. Zeit., xvi, 1872.
Loew, H. Centuries of Descriptions North American Diptera. (Berlin Ent Zeit.)
Macquart, J. New or little-known exotic Diptera. 9 vols., 8vo. Paris, 1838-55. (Fr.)
Macquart, J. Natural History of the Diptera. (Suites à Buffon). 2 vols., 8vo. Paris, 1834-5. (Fr.)

Macquart, J. Diptera of the North of France. (5 parts.) 8vo. Lille, 1826-33. (Fr.)

Meigen, J. W. Classification and Description of European Diptera. 4to. 1804. (Ger.)

Meigen, J. W. Systematic descriptions of the known European Diptera. 7 vols., 8vo. Aix and Hamburg, 1818-1838.

Meigen, J. W. Same continued by Loew, vols. 8-10. Halle, 1869-73.

Osten Sacken, C. R. Catalogue of Described Diptera of North America. (Smithsonian Misc. Coll., iii.) Washington, 1859.

Osten Sacken, C. R. New N. A. Tipulidæ with short palpi. (Proc. Phila. Acad., 1860.)

Osten Sacken, C. R. Diptera in Wheeler's Survey. Zoology. 4to, 1875.

Osten Sacken, C. R. North American species of Syrphus. (Proc. Bost. Soc., xviii, 1875.)

Osten Sacken, C. R. List of North American Syrphidæ. (Bull. Buff. Soc.) 1877.

Osten Sacken, C. R. Prodrome of a Monograph of the Tabanidæ of the U. S. (Mem. Bost. Soc.) 1875-6.

Robineau Desvoidy, J. B. Natural History of the Diptera of the neighborhood of Paris. 8vo. Paris, 1863. (Fr.)

Rondani, C. Prodrome of Italian Dipterology. 8vols., 8vo. Parma, 1856-77. (Lat.)

Schellenberg, J. R. Genera of Flies. 8vo, 42 pls. Zurich, 1803.

Schiner, J. R. Dipterous Fauna of Austria. 2 vols., 8vo. Vienna, 1860-64. (Ger.)

Schiner, J. R. Diptera of the voyage of the Novara. 4to. Vienna, 1868. (Ger.)

Siebke et Schneider. Enumeration of Norwegian Insects. iv, Diptera. 8vo. Christiania, 1877.

Walker, F. List of the Diptera in the Collection of the British Museum. 7 parts, 12 mo. 1848-55.

Walker, F. Insecta Brittanica; Diptera. 3 vols., 8vo. London, 1851-56.

Walker, F. Catalogues of Diptera collected by Wallace in the East Indies. 8 parts. London, 1856-64.

Werdemann, C. R. Diptera exotica. 1 vol., 12mo. 1821.

Werdemann, C. R. Extra European Diptera. 4 vols. 8vo. 1828-30. (Ger.)

Winnerz, J. Monograph of the Mycetophilidæ. (Verhandlung Zoöl. Bot. Soc.) 8vo. Vienna, 1863.

Zetterstedt, J. W. Scandinavian Diptera arranged and described. 14 vols., 8vo. London, 1842-60.

COLEOPTERA.

Austin, E. P. Species of Sunius and Pæderus in U. S. (Proc. Bost. Soc., xviii, 1876.)

Baly, J. S. Descriptive Catalogue of Hispidæ. 8vo. London, 1858.

Bland, J. H. Descriptions of Staphylinidæ of U. S. 8vo. Phila., 1865.

Bohemann, C. H. Monograph of Cassididæ. 4 vols., 8vo. Stockholm, 1850-62. (Lat.)

Bohemann, C. H. Catalogue of Cassidæ in British Museum. 8vo. London, 1856.

Boisduval. Coleoptera of the Voyage of the Astrolabe. 8vo. Paris, 1835. (Fr.)

Bonvouloir, H. Monograph of Throscidæ. 8vo. Paris, 1859-60. (Fr.)

Bonvouloir, H. Monograph of Eucnemidæ. 8vo, 42 pls. Paris, 1871-5. (Fr.)

Brême, F. Monograph of Cossyphidæ. 8vo and 4to. Paris, 1842-6. (Fr.)

Candeze, E. Monograph of Elateridæ. 4 vols. and suppl. Liège, 1857-64. (Fr.)
Castelnau and Gory. Natural History and Iconography of Coleoptera. 6 vols., 8vo, 260 pls. Paris, 1835-41. (Fr.)
Chapnis, J. Monograph of Platypidæ. 8vo. Liège, 1866. (Fr.)
Chenu. Encyclopedia of Natural History. Coleoptera. 3 vols., 4to, 1800 figs. Paris, 1851-61.
Chevrolat, A. Coleoptera of Mexico. 8vo. Strasburg, 1834-42. (Fr.)
Clark, H. Catalogue of Halticidæ in British Museum. Pt. 1. 8vo. London, 1860.
Cox, H. E. Handbook of Coleoptera of Gt. Britain. 2 vols., 8vo. London, 1873.
Crotch, G. R. Materials for study of Phytophaga of U. S. (Proc. Phila. Acad., 1873)
Crotch, G. R. Check List of North American Coleoptera. 8vo. Salem, 1874.
Crotch, G. R. Revision of the Coccinellidæ and Erotylidæ. 8vo. London, 1874-6.
Crotch, G. R. Synopses Erotylidæ, Endomychidæ, Coccinellidæ and Dytiscidæ of U. S. (Trans. Am. Ent. Soc., iv, 1873)
Dejean et Aubé. Species Generale de Coléoptères 6 vols., 8vo. Paris, 1825-38. (Fr.)
Dejean, Boisduval et Aubé. Iconographic Natural History of the Coleoptera of Europe. 4 vols , 8vo, 223 pls. Paris, 1829-34 (Fr.)
Duval et Fairmaire. Genera of European Coleoptera. 4 vols., 8vo, 300 pls. Paris, 1857-68. (Fr.)
Erichson, W. F. Genera and species of Staphylinidæ. 8vo Berlin, 1840. (Lat)

Erichson, W. F. Conspectus of Peruvian Coleoptera. 8vo. Berlin, 1847. (Lat.)

Fairmaire et Germaine. Revision of Coleoptera of Chili. 8vo. Paris, 1860-63. (Fr.)

Gemminger et Harold. Catalogue of all described coleoptera with synonyma. 11 vols., 8vo. Munich, 1868-74. (Lat.)

Girard, M. Treatise on Entomology. Coleoptera. 8vo, 60 pls. Paris, 1873. (Fr.)

Gory et Percheron. Monograph of Cetonias and allied Genera. 8vo, 77 pls. Paris, 1833. (Fr.)

Haldeman, S. S. North American Cryptocephali. 4to. Phila., 1849.

Haldeman, S. S. Materials towards a Monograph of the Coleoptera Longicornia of the U.S. 4to. Phila., 1847.

Hope, F. W. Coleopterist's Manual. 3 vols., 8vo. London, 1837-40.

Hope, F. W. Buprestidæ of New Holland. 8vo.

Hope, F. W. Monograph of Genus Euchlora. (Ann. Nat. Hist., iv, 1840.)

Horn, G. H. Revision of North American Tenebrionidæ. (Trans. Am. Phila. Soc., 1870.)

Horn, G. H. Revision of species of Acmæodera of U. S. (Trans. Am. Ent. Soc., vii, 1878.)

Horn, G. H. Synopsis of Parnidæ. Trans. Am. Ent. Soc., ii, 1870.

Horn, G. H. Contributions to the Coleopterology of the U. S. Trans. Am. Ent. Soc., ii, 1870.

Horn, G. H. Descriptive catalogue of species of Nebria and Pelophila. Trans. Am. Ent. Soc., ii, 1870.

Horn, G. H. On the species of Oödes and allied genera of U. S. Trans. Am. Ent. Soc. ii, 1870.

Horn, G. H. Descriptions of the species of Aphoderus and Dialytes. Trans. Am. Ent. Soc., ii, 1870.

Horn, G. H. Synopses of Corphyra and Aphodiini of U. S. Trans. Am. Ent. Soc., ii, 1871.

Horn, G. H. Synopsis of Malachidæ of U. S. Trans. Am. Ent. Soc , iv, 1872.

Horn, G. H. Brenthidæ of U.S. Trans. Am. Ent. Soc., iv, 1872.

Horn, G. H. Revision of species of Lebia. Trans. Am. Ent. Soc., iv, 1872.

Horn, G. H. Contributions to a knowledge of the Curculionidæ of the U. S. Proc. Am. Phil. Soc., 1873.

Horn, G. H. Revision of Meloidæ, Hydrobiini, Histeridæ and Bruchiqæ. Proc. Am. Phil. Soc., 1873.

(For a full list of Papers on Coleoptera by George H. Horn, see *Psyche* for 1878.)

Lacordaire, T. Genera of Coleoptera, continued by Chapius. 12 vols. 8vo with atlas, 134 plates. Paris, 1854-76. (Fr.)

Lacordaire, T. Revision of the Cicindelidæ. 8vo, Paris, 1843. (Fr.)

Lacordaire, T. Monograph of the Phytophagous Coleoptera. 2 vols. 8vo, Paris, 1845-8. (Fr.)

Leconte, J. L. Synonymical Note on N. A. Coleoptera (a review of the types of Kirby, Newman, Walker and those of the Parisian collections). Ann. and Mag. Nat. Hist., 1870.

Leconte, J. L. Classification of the Coleoptera of the U. S. Smithsonian Misc. Coll., iii and xi. 1861-73.

Leconte, J. L. Classification of the Rhynchophorous Coleoptera. Amer. Naturalist, viii, 1874.

Leconte, J. L. New species of N. A. Coleoptera. Smithsonian Misc. Coll., 1863-1873.

Leconte, J. L. List of North American Coleoptera. Smith. Misc. Coll., 1863 ———.

Leconte, J. L. Coleoptera of Kansas and Eastern New Mexico. (Smithsonian Contributions xi, 1859

Leconte, J. L. Pterostichi of the U. S. (Jour. Phila. Acad. ii, 1852) Revision. Proc. Phila. Acad., 1873.

Leconte, J. L. Synomymical remarks on N. A. Coleoptera. Proc. Phil. Acad., 1873.

Leconte, J. L. Revision of the Cicindelidæ of U. S. (Phila. 4to, 1856).

Leconte, J. L. Synopsis of Melolonthidæ of U. S. (Phila. 4to, 1856).

Leconte, J. L. Revision of Buprestidæ. (Phila. 4to, 1859.)

Leconte, J. L. Attempt to classify the Longicorn Coleoptera of N. A. (Jour. Phil. Acad. ii, 1853.)

Leconte, J. L. Coleoptera in Agassiz's Lake Superior. 1850.

Leconte, J. L. Revision of the Elateridæ of the U. S. (Phila.) 4to, 1853.

Leconte, J. L. Notes on Classification of Carabidæ of U. S. 4to (Phila.), 1853.

(For a full list of Papers on Coleoptera by John L. Leconte, see Psyche, 1878.)

Marseul, S. A. Monograph of the Histeridæ. 2 vols. 8vo, Paris, 1853-59. (Fr.)

Matthews, A. Trichopterygia illustrata et descripta. 8vo, 31 plates, London, 1871.

Melsheimer, F. E. Catalogue of North American Coleoptera. Edited by Haldeman and Leconte. Smithsonian Misc. Coll., 1853.

Mulsant, E. Natural History of the Coleoptera of France. 27 vols. 8vo, many plates. Paris, 1839-75. (Fr.)

Olivier, A. G. Natural History of Insects. Coleoptera. 6 vols. 4to, atlas, 363 pls. Paris, 1789-1808. (Fr.)

Pascoe, F. P. Longicornia of Australia. 8vo, London, 1867.

Putzeys, J. Revision of the Clivinidæ. 8vo. Brussels, 1867-8. (Fr.)

Redtenbacher, L. Austrian Coleoptera. 3rd edit. Vienna. 71-74. (Ger.)

Saunders, E. Catalogus Buprestidarum. 8vo. London, 1871.

Schaufuss, L. W. Monograph of genus Machærites. Verh. Z. B. Gesell. Wien, 1863.

Schiodte. De Metamorphosi Eleutheratorum observationes. (Naturhistorisk Tidsskrift. 1861-77. 72 pls. Lat.)

Schönherr, C. J. Genera et Species Curculionidum. 8 vols. 8vo. Paris, 1833-45.

Sharp, D. Coleoptera of the Amazon Valley. Staphylinidæ. (Lond.) 1876.

Spinola, M. Monograph of the Cleridæ. 2 vols. 8vo. 47 pls. 1844 (Fr.)

Sprague, P S. Notes on some of the common species of Carabidæ of North America. Canadian Entomologist, ii, 1870.

Spry and Shuckard. The British Coleoptera delineated. 8vo. 93 pls. Lond., 1861.

Stal, C. Monograph of American Chrysomelidæ. 4to, Upsala, 1862-5. (Fr.)

Strauch, A. Systematic catalogue of Coleoptera described in the Annales de la Société Entomologique de France. 8vo. Halle, 1861.

Sturm, J. German Coleoptera. 23 vols. 8vo. 424 plates. Nurenburg, 1805-1857. (Ger.)

Thompson, J. Monograph of the Cicindelidæ. fol. Paris, 1859. (Fr.)

Thompson, J. Classification of the Cerambycidæ. 8vo. Paris, 1860. (Fr.)

Thompson, J. Systema Cerambycidarum. 4 pts. 8vo. Paris. 1866-68.

Thompson, C. G. Scandinavian Coleoptera. Synoptically arranged. 10 vols. 8vo. Lund., 1859-69.

Ulke, H. Coleoptera in Report Wheeler's Survey, v. 4to, Washington, 1875.

Walton, John. List of British Curculionidæ. London, 1856.

White, Smith and Boheman. Catalogue of Coleoptera in British Museum. 9 pts. pub. 12mo. Lond., 1847-56.

Zimmermann. Synopsis of Scolytidæ of North America. 8vo. Phila., 1868.

HEMIPTERA.

Amyot, C. J. B. Entomologie Francais — Rhynchotes. 8vo. Paris, 1848. (Fr.)

Amyot et Serville. Natural History of the Hemiptera (suites à Buffon). 1 vol. 8vo. Paris, 1843. (Fr.)

Bürensprung, F. Catalogue of European Hemiptera. 8vo. Berlin, 1860. (Ger.)

Buckton, G. B. Monograph of the British Aphides. (Ray Society). 2 vols. 8vo. London, 1876-79.

Dallas, W. S. List of the Hemiptera in British Museum. 2 pts. 8vo, London, 1851-2.

Delaporte, F. L. Systematic classification of the Hemiptera. 8vo. Paris, 1833. (Fr.)

Denney. Monographia Anoplurum Brittaniæ. (British Lice.) 1 vol. 8vo. London, 1842.

Dohrn, A. Catalogue of Hemiptera. 8vo. 1859.

Douglas and Scott. British Hemiptera Vol. 1. Heteroptera. (Ray Society.) 8vo. London, 1865.

Fieber, F. X. European Hemiptera. 8vo. Vienna, 1861. (Ger.)

Giebel, C. G. Insecta Epizoa Parasites of Mammals and Birds. fol. 20 pls. Leipzig, 1874. (Ger.)

Glover, T. Manuscript Notes: Hemiptera. **4to 10 pls.** Washington, 1876.

Hahn und Herrich Schäffer. The Hemipterous Insects. 9 vols. 8vo. 324 pls. Nurenburg, 1831-53. (Ger.)

Herrich Schäffer. Synonymical Index of Heteropterous Hemiptera. 8vo. Regensburg, 1853.

Koch, C. L. The Plant Lice figured and described. 8vo 54 plates. Nurenburg, 1857. (Ger.)

Mulsant et Rey. Natural History of the Bugs of France. 4 vols. 8vo. Lyons, 1865-73.

Reuter, O. M. American Acanthidæ. (Roy. Swed. Acad.) 8vo. 1871. (Lat.)

Reuter, O. M. North American Cupsinæ (l. c.), 1875. (Lat.) (See also other papers published by the Swedish Academy.)

Riley and Monell. Notes on Aphidæ of the U. S. (Bull. U. S. Geol. Surv., v.) 8vo, Washington, 1879.

Saunders, E. Synopsis of British Heteroptera 8vo. London, 1875.

Say, T. Descriptions of new species of Heteropterous Hemiptera. (Reprinted Trans. N. Y. State Agr. Soc., 1857.)

Siebke, H. Enumeratio Insectorum Norvegicum. I. Hemiptera. 8vo. Christiania, 1874.

Stal, C. Papers published by Royal Swedish Academy. 1853.

Th mas, C. List of the Aphidini of the U. S. (Bulletin Illinois Indust. Univ. No. 2. 1878.)

Thomas, C. Monograph of the Plant Lice. (8th Report State Entomologist of Ill , 1879.)

Uhler, P. R. Notices of Hemiptera of Western Territories. (Rep. U. S. Geol. Surv., 1871). 8vo, 1872.

Uhler, P.R. Monographs Cydnidæ and Saldæ [etc.]. (Bull. U. S. Geol. Surv. iii, 1877).

Uhler, P. R. Hemiptera of Wheeler's Survey—Zoölogy. v, 1875.

Walker, F. List of Homoptera in British Museum. 5 parts. 8vo. London, 1850-58.

Walker, F. Catalogue of Hemiptera Heteroptera in British Museum. 8 vols. and suppl. 8vo. 1867-74.

ORTHOPTERA.

Brunner, C. Monograph of the Phaneropteridæ. 8vo. Vienna, 1878. (Ger.)

Brunner, C. Orthopterological Studies (Verh. Z. B. Gesellschaft Wien). 8vo. 1861. (Ger.)

Charpentier, T. de. Orthoptera figured and described. 1 vol., 4to, 60 pls. Leipzig, 1841-5. (Lat.)

Fischer, L. H. Orthoptera Europæa. 4to. Leipzig, 1853.

Fischer, de Waldheim. Orthoptera Imperii Rossici (cont. by Eversmann). 4to. Moscow, 1846-7. (Lat.)

Gray, G. R. Entomology of Australia. Pt. 1, Monograph of the genus Phasma (all pub.) 4to, London, 1833.

Gray, G. R. Synopsis of the Family Phasmidæ. 8vo. London, 1835.

Glover, T. Illustrations of North American Entomology. Orthoptera. 4to, 13 pls. Washington, 1872.

Saussure, H. Natural History of the Orthoptera of Central America, Mexico and the Antilles. Blattariæ. 2 parts, 4to. Geneva, 1862-5. (Fr.)

Saussure, H. Orthoptera of Mexico: Mantides. 4to. Geneva, 1871. (Fr.)

Saussure, H. Orthopterological Miscellany. 6 pts., 4to. Geneva, 1863-77. (Fr.)

Saussure, H. Studies of the Orthoptera of Mexico and Central America. (3 pts. pub.) Folio. Paris, 1870-74. (Fr.)

Scudder, S. H. Materials for a Monograph of the N. A. Orthoptera. (Jour. Bost. Socy., vii). 8vo. 1862.

Scudder, S. H. Catalogue of North American Orthoptera. (Smithsonian Misc. Coll., viii). 8vo. Washington, 1868.

Scudder, S. H. On the arrangement of the Families of Orthoptera. (Proc. Bost. Socy., xii). 8vo. Boston, 1869.

Scudder, S. H. Revision of the Mole Crickets (Mem. Peabody Acad. No. 1). 4to. Salem, 1869.

BIBLIOGRAPHY. 197

Scudder, S. H. Notes on Forficulariæ with list of described species. (Proc. Bost. Socy., xviii). 8vo. 1876.

Scudder, S. H. A Century of Orthoptera (Proc. Bost. Socy.).

Scudder, S. H. Remarks on Calliptenus and Melanophus with a notice of New England species (Proc. Bost. Socy., xix). 1877.

Serville, A. Natural History of the Orthoptera (suites à Buffon). 1 vol., 8vo. Paris, 1839. (Fr.)

Siebke, H. Enumeratio Insectorum Norvegicum: I, Orthoptera. 8vo. Christiania, 1874.

Stal, C. Recensio Orthopterorum. (Review of Orthoptera described by Linne, De Geer, etc.) 3 pts, 8vo. Stockholm, 1873-5.

Stoll, C. Representations of the Phasmidæ, Mantidæ, Acrididæ, Gryllidæ (etc.), of the four parts of the world. 2 vols. 4to. 70 pls. Amsterdam, 1815. (Fr.)

Thomas, C. List and description of new species of Orthoptera (Rep. U. S. Geol. Surv., 1870). 8vo. Washington, 1871.

Thomas, C. Synopsis of Acrididæ of North America. (Final Report U. S. Geol. Surv., Vol. v). 4to. Washington, 1875.

Thomas, C. Orthoptera in Rep. Wheeler's Survey, v. 1875.

Walker, F. Complete Catalogue of genera and species of Blattariæ. 8vo. London, 1878.

Walker, F. Complete Catalogue of Dermaptera Saltatoria. 4 vols. and suppl., 8vo. London, 1869-71.

Westwood, J. O. Catalogue Orthoptera in British Museum: I, Phasmida (all pub.) 4to, 48 pls. London, 1859.

NEUROPTERA.

Charpentier. European Libellulina figured and described. 4to, 48 pls. Leipzig, 1840. (Lat.)

Eaton, A. E. Monograph Ephemeridæ. Pt. 1. (Trans. Ent. Socy., London.) 8vo. 1871.

Hagen, H. A. Special Monograph of the Termites. 4 pts., 8vo. (Berlin) 1858. (Ger.)

Hagen. H. A. Catalogue of Neuroptera in British Museum: Termetina. 8vo. London, 1858.

Hagen, H. A. Synopsis of described Neuroptera of N. A. (Smithsonian Misc. Coll , iv). 8vo. Washington, 1861.

Hagen, H. A. The Larvæ of the Ant Lions. (Verh. Entom. Vereins Stettin, xxxiv.) 8vo. 1873. (Ger.)

Hagen, S. H. Neuroptera collected by Lieut. Carpenter (Rep. U. S. Geol. Surv., 1873.) 8vo. 1874.

Hagen, H. A. Neuroptera in Wheeler's Survey, v. 4to, 1875

Hagen, H. A. Synopsis of the Odonata of America. (Proc. Bost. Socy., xviii.) 8vo. Boston, 1875.

Hagen, H. A. Pseudoneuroptera and Neuroptera in collection of T. W. Harris. (Proc. Bost. Soc., xv, 1873).

Kolenati, F. A. Genera and species of Trichoptera. 2 pts., 4to. Prague, 1848. Moscow, 1859. (Lat.)

McLachlan, R. Monograph British Psocidæ. 8vo. London, 1867.

McLachlan, R. Monographic revision of the Trichoptera of Europe, incomplete. Parts 1-7. 44 pls., 8vo. London, 1874-8.

Newport, G. On Pteronarcys regalis with descriptions of American Perlidæ. 4to. (London.) 1851.

Pictet, F. J. Natural History of the Neuroptera. 2 vols., 8vo., 100 pls. Geneva, 1841-3. (Fr.)

Rambur, P. Natural History of the Neuroptera (suites à Buffon.) 1 vol., 8vo. Paris, 1842. (Fr.)

De Selys Longchamps, E. Monograph of the Libellulæ of Europe. 8vo. Paris, 1840. (Fr.)

De Selys Longchamps and Hagen. Review of the Odonata of Europe. 8vo. Brussels, 1850. (Fr.)

De Selys Longchamps and Hagen. Monographs of the Calopteryginæ and Gomphinæ. 8vo. 37 pls. Brussels, 1854 and 58. (Fr.)

Walker, F. Catalogue of Neuroptera in British Museum. 4pts. and suppl. 12mo. London, 1852-8.

SYMPHILA.
(*Thysanura* and *Collembola*.)

Lubbock, J. Notes on the Thysanura. (Trans. Linn. Soc.) 4to. 7 pls. London, 1867-70.

Lubbock, J. Monograph of the Collembola and Thysanura. (Ray Soc.) 8vo. 78 pls. London, 1873.

Meinert, F. Campodeæ, a Family of Thysanura. (Naturhist. Tidskrift) 8vo. Copenhagen, 1865. (Swedish and Latin.)

Nicolet, H. Researches on the Podurellas. 4to. 9 pls. Neuchatel, 1842. (Fr.)

Packard, A. S., jr. Bristle Tails and Spring Tails. (American Naturalist) 8vo. Salem, 1871.

Packard, A. S., jr. Synopsis of Thysanura of Essex Co., Mass. (5th Report Peabody Academy) 8vo. Salem, 1872.

Tullburg, T. F. Swedish Poduridæ. 8vo. 12 pls. Stockholm, 1872.

Tullburg, T. F. Collembola boreali. (Roy. Swedish Acad.) 8vo. 1876.

ARACHNIDA.

Ausserer, A. Contributions to the Territellariæ. (Verh. Zoöl. Bot. Ges., Wien, XXI, 1871. (Ger.)

Blackwall, J. Spiders of Great Britain and Ireland. (Ray Socy.) 2 vols. 4to. 29 pls. London, 1861-66.

Cambridge, O. P. Papers in Proceedings of various Societies of London.

Doleschall, C. L. Contributions to a knowledge of the Arachnida of the East Indies. 2 pts. Batavia, 1856-9. (Dutch.)

Furstenberg, H. The Itch Mites of Men and Animals. fol. 15 pls. Leipzig, 1861. (Ger.)

Hahn, C. W. Monographia aranearum. 8 pts. 4to. 32 plates, Nurenburg, 1820-36. (Ger.)

Hahn und Koch. Die Arachniden. 16 vols. 8vo. 560 plates. Nurenburg, 1831-48. (Ger.)

Haldemann, S. S. On some species of Hydrachnidæ. (Zoölogical Contributions No. 1.) 8vo. Phila. 1842.

Hentz, N. M. Spiders of the United States. (Jour. Bost. Soc. vol. 1-6 and Proc. ii).

Hentz, N. M. Reprint by Emerton and Burgess. 1 vol. 8vo. (Occas. Papers Bost. Soc.) 21 plates, 1875.

Herbst, J. F. W. Natural History of Apterous Insects. 4 parts. 4to. 23 plates, Berlin, 1797-1800. (Ger.)

Hodge, G. On British Pycnogonida. (Annals and Mag. Nat. Hist.) London, 1864.

Johnston, Geo. An attempt to ascertain the British Pycnogonida. (Mag. Zool. and Bot. i) London, 1837.

Keyserling. Papers in Verh. Zoöl. Bot. Ges. Wien, 1862. (Ger.)

Koch, C. L. Review of the Arachnida. 8vo. 49 pls. Nurenberg, 1837-50. (Ger.)

Koch, C. L. Arachnida of the Family Drassidæ. 7 parts (all pub.) 8vo. Nurenburg, 1866-8. (Ger.)

Koch, C. L. Australian Arachnida. (22 parts pub.) 4to. Nurenburg, 1871-8. (Ger.)

Kramer, P. Systematic Revision of the Mites (Archiv für Naturgeschichte xliii) 8vo. 1877. (Ger.)

Kramer, P. Natural History of Hydrachnidæ. (Archiv für Naturgesch., 1875.

Menge, A. Prussian Spiders. 8vo. 81 pls. Dantzig, 1866-76.

Nicolet, H. Acarina of the environs of Paris. 4to. 10 pls. Paris, 1855. (Fr.)

Say, T. Description of the Arachnides of the U.S. (Jour. Phila. Acad., 1821)

Simon, E. Natural History of the Araneides. 8vo. Paris, 1864. (Fr.)

Simon, E. The Arachnida of France. (4 vols. pub.) 8vo. Paris, 1874-8. (Fr.)

Stanley, E. J. British Spiders. 8vo. London, 1866.

Thorell, T. European Spiders (1 pt. pub.) 4to. Upsala, 1869-70. (English.)

Thorell, T. Spiders from Labrador. (Proc. Bost. Soc., 1875.)

Thorell, T. Studies of Malay and Papuan Spiders. 8vo. Genoa, 1877. (Lat.)

Thorell, T. Classification of Scorpion. (Am. and Mag. N. H. Jan., 1876.) (A list of the literature of European spiders will be found in Thorell's European Spiders *supra*.)

Vinson, A. Arachnida of Reunion, Mauritius and Madagascar. 4to. Paris, 1863. (Fr.)

Walckener et Gervais. Natural History of the Aptera (suites à Buffon). 4vols. 8vo. 52 pls. Paris, 1837-47. (Fr.)

Westring, N. Swedish Arachnida. 8vo. Gothenburg, 1862. (Lat.)

Wilson, E. B. Synopsis of Pycnogonida of New England. (Trans. Conn. Acad., v, 1878.)

Wilson, E. B. The Pycnogonida of New England and adjacent waters. Rep. U. S. Fish Comm., 1881, pp. 43, pls. 7.

Wood, H. C. Phalangeæ of the U. S. (Proc. Essex Inst., vi) 8vo. Salem, 1868.

Wood, H. C. Pedipalpi of North America. (Trans. Am. Phila. Soc.) 4to. 1863.

MYRIAPODA.

Cope, E. D. New or little known Myriapoda from South Western Virginia. (Proc. Am. Phil. Soc.) 1869.

Harger. New North American Myriapoda. (Am. Jour. Sci. iv) 8vo. New Haven, 1874.

Humbert, A. Myriapods of Ceylon. 4to. Geneva, 1865. (Fr.)

Koch, C. L. The Myriapods. 2 vols. 8vo. 119 pls. Halle, 1863. (Ger.)

Meinert, F. Myriapods of the Copenhagen Museum. 8vo. Cogenhagen, 1812. (Lat.)

Newport, Geo. Monograph of the Myriapoda Chilopoda. (Trans. Linn. Soc. xix) 4to. London, 1844-5.

Saussure, H. Myriapod Fauna of Mexico. 4to. 7pls. Geneva, 1860. (Fr.)

Saussure et Humbert. Studies of the Myriapods of Mexico and Central America with a catalogue of American Myriapoda. fol. Paris, 1872. (Fr.)

Stuxberg, A. North American Lithobidæ and Oniscidæ. (Forh. Roy. Sw. Acad.) 8vo. Stockholm, 1875. (Swedish.)

Wood, H. C. Myriapoda of North America. (Trans. Am. Phil. Soc.) 4to. 1865.

CRUSTACEA.

Baird, W. History of the British Entomostraca (Ray Soc.) 8vo, 36 pls. London, 1850.

Bate, C. S. Catalogue of Amphipodous Crustacea in British Museum. 1 vol. 8vo. 59 pls. London, 1862.

Bate and Westwood. British Sessile-eyed Crustacea. 2 vols. 8vo. London, 1863-9.

Bell, T. Monograph of the Leucosoidea. (Trans. Linn. Soc. xxi.) 4to. London, 1855.

Birge, E. A. Notes on Cladocera. (Trans. Wisconsin Acad. iv.) 1878.

Boeck, A. Boreal and Arctic Amphipod Crustacea. (Vidensk. selsk. Forh.) 8vo. 1870. (Latin.)

Boeck, A. Contributions to Californian Amphipod Fauna (l.c.) 8vo. 1871.

Brady, G. S. Monograph of recent British Ostracoda. (Trans. Linn. Socy., xxvi.) 4to. 19 pls. 1868.

Brady, G. S. Monograph of free and semiparasitic Copepoda of British Isles. (Ray Socy.) 8vo. 1878.

Brandt, J. F. Crustacea in Middendorff's Siberian Journey. 4to. St. Petersburg, 1851.

Claus, C. Free Copepoda of North Sea and Mediterranean. 4to. 37 pls. Leipzig, 1863.

Dana, J. D. Crustacea of the U. S. Exploring Expedition. 2 vols. 4to. and folio atlas. 96 pls. Phila., 1852.

(Preliminary papers were published in Proc. Phila. Acad., Proc. Am. Acad., and Am. Jour. Sci.)

Darwin, C. Monograph of Cirripedia. (Ray Society.) 2 vols. 8vo. 40 pls. London, 1851-4.

DeKay, J. E. N. Y. Fauna: Crustacea. 4to. 13 pls. Albany, 1844.

Gerstaecker, A. Carcinological Contributions (Archiv für Naturgeschichte xxi) 8vo. 1856. (Ger.)

Gerstaecker, A. Classes and orders of Arthropoda in Brown's "Klassen und Ordnungen" 1866 — (incomplete.)

Gibbes, L. R. On the Carcinological Collections of the U. S. (Proc. Am. Assoc. iii, 1851).

Gamroth. Contributions to a Natural History of the Caprellidæ. (Zeitsch. Wiss. Zoologie, xxviii, 1878) (Ger.)

Hagen, H. A. Monograph N. A. Astacidæ. (Ill. Catalogue Mus. Comp. Zool) 4to. Cambridge, 1870.

Harger, O. Description of new genera and species New England Isopoda. (Am. Jour. Sci. xv, 1878.)

Harger, O. Notes on New England Isopoda. (Bull. U. S. Nat. Mus., 1879.)

Harger, O. Marine Isopoda of New England. Rep. U. S. Fish Com., 1878 [1880].

Heller, C. Crustacea of Southern Europe. 8vo. Vienna, 1863. (Ger.)

Heller, C. Crustacea of the voyage of the Novara. 4to, 25 pls. Vienna, 1868. (Ger.)

Heller, C. Contributions to the Crustacean Fauna of the Red Sea. (Stzb. K. K. Akad Wien. xliii, i, 1861.)

Herbst, J. F. W. Natural History of Crabs and Crawfish. 3 vols. 4to. Atlas. 62 plates. fol. Berlin, 1790-1804.

Hess, W. Contributions to a knowledge of the Decapoda of East Australia. (Archiv für Naturgesch. 1865.) (Ger.)

Hilgendorf, F. Crustacea in van der Decken's Travels in East Africa. 8vo. Leipzig, 1869. (Ger.)

Kingsley, J. S. Synopsis of North American Alphei. (Bull. U. S. Geol. Surv. iv, 1878.)

Kingsley, J. S. Crustacea from Virginia and Florida with a revision of the genera of Caridea. (Proc. Phil. Acad., 1879.)

Kingsley, J. S. Carcinological Notes, Revision of Gelasimus, Ocypoda and Grapsidæ. (Proc. Phil. Acad., 1880).

Kossmann, R. On the Rhizocephala. Arbeiten Zool. Zoot. Inst. Wurzburg i, 1874. (Ger.)

Kosmann, R. Zoological Results of Travels in the neighborhood of the Red Sea. 4to. Leipzig, 1877. (Ger.)

Krauss, F. South African Crustacea. 4to. Stuttgart, 1843. (Ger.)

Kroyer, H. Greenland Amphipoda. 4to. Copenhagen, 1837. (Danish.)

Kroyer, H. Monograph of Northern Hippolytes. 4to. Copenhagen, 1842. (Danish.)

Kroyer, H. Monograph of Sergestes. 4to. Copenhagen, 1856. (Danish.)

Kroyer, H. Contributions to a Knowledge of the Parasitic Crustacea. 8vo, 18 pls. Copenhagen, 1863. (Danish.)

Leach, W. E. British stalk-eyed Crustacea, continued by Sowerby. 19 parts, 4to, 54 pls. London, 1815-1875.

Lereboullet, A. Oniscidæ of Alsace. 4to, 10 pls.. Strasburg, 1853. (Fr)

Leydig, F. Natural History of the Daphnida. 1 vol., 4to. Tübingen, 1860. (Ger.)

Lilljeborg, Crustacea, Cladocera, Ostracoda et Copepoda Scaniæ. 4to, 27 pls. Lund., 1853. (Lat.)

Lockington, W. N. Papers in Proceedings California Acad. and Annals and Mag. Nat. Hist., 1874. (Pacific forms.)

Lutken, C. F. Contributions to a Knowledge of the genus Cyamus. 4to. Copenhagen, 1873. (Danish.)

von Martens, E. On Cuban Crustacea. (Archiv für Naturgesch, xxxviii). 8vo., 1872. (Ger.)

Meinert, F. Crustacea Isopode Amphipoda et Decapoda Daniæ. (Naturhist. Tidsskrift iii, xi.) 8vo. 1877.

Miers, E. J. Notes on oxystomatous Crustacea. (Trans. Linn. Soc., 1876.)

Miers, E. J. Crustacea from the Malaysian Region. (Ann. and Mag. N. H., V. v.) 8vo. London, 1880.

Miers, E. J. Notes on Peneidæ. (Proc. Zool. Socy. London, 1878.)

Miers, E. J. Revision of the Hippidea. (Jour. Linn. Socy. 1878.)

Miers, E. J. On the Classification of the Maioid Crustacea. (Jour. Linn. Socy. xiv. 1879.)

Miers, E. J. Revision of the Plagusinæ. (Ann. and Mag. N. H. 1878.)

Miers, E. J. On the Squillidæ. (Ann. and Mag. N. H., vol. v. 1880.)

Miers, E. J. Catalogue of New Zealand Crustacea. 8vo. London, 1876.

Milne-Edwards, A. Studies of the Recent Portunidæ. (Archives du Museum, x, 1861.) 4to, 11 pls. Paris, 1861. (Fr.)

Milne-Edwards, A. Researches on the Carcinological fauna of New Caledonia. (Nouv. Arch. viii & ix.) 4to. Paris, 1872-3. (Fr.)

Milne-Edwards, A. Revision of the genus Thelphusæ. (Nouv. Arch. v, 1869.) (Fr.)

Milne-Edwards, A. **Studies of the Cancridæ.** (Nouv. Arch. i.) 4to. 1866. (Fr.)

Milne-Edwards, H. **Crustacea of Mission scientifique au** Mexique et dans Amerique meridionale (incomplete). 4to, many plates. Paris, 1873. (Fr.)

Milne-Edwards, H. **Natural History of the Crustacea** (suites à Buffon). 3 vols. and atlas, 8vo. Paris, 1837-42. (Fr.)

Milne-Edwards, H. Notes on new or little-known Crustacea. (Archives Museum vii.) 4to. 1854.

Milne-Edwards and Lucas. Crustacea in D'Orbigny's Voyage to Central America. 4to, 17 pls. Paris, 1844. (Fr.)

Müller, Fritz. Facts for Darwin, translated by Dallas. 8vo. London, 1869.

Müller, P. E. Denmark's Cladocera. 8vo. Copenhagen, 1868.

Ordway, A. Monograph of the genus Callinectes. (Jour. Bost. Socy., 1863.)

Owen, R. Crustacea of Beechey's Voyage of the Blossom. 4to. London, 1839.

Packard, A. S. jr. **Synopsis of fresh-water Phyllopoda of N. A.** (Rep. U. S. Geol. Survey, 1873 [1874].)

Randall, J. W. Crustacea from the West Coast of North America. (Jour. Phila. Acad. viii, 1839).

Rüppell, E. Crabs of the Red sea. 4to. Frankfort, 1830. (Ger.)

Sars, G. O. Review of Norse Marine Ostracoda. 8vo. Christiania, 1865. (N.)

Sars, G. O. Natural History of fresh-water Crustacea of Norway. 4to. Christiania, 1867. (Fr.)

Sars, G. O. Monograph of Mysidæ of Norway. 4to. Christiania, 1870-2. (N.)

Sars, G. O. Cumacea Oceani Atlantici. 4to, 20 pls. Stockholm, 1871.

Sars, M. Contributions to a Knowledge of the Fauna of Christiania fiord. 3 pts., 8vo. (Mollusca & Crust.) Christiania, 1868–73. (N.)

Saussure, H. Various new Crustacea from the Antilles and Mexico. (Mem. Soc Phys. et d'Hist. Nat. Genève, xiv). 4to. Geneva, 1858. (Fr.)

Saussure, H. New Crustacea from the West Coast of Mexico. (Rev. et Mag. Zool. II, v, 1853.) (Fr.)

Say, T. An Account of the Crustacea of the U. S. (Jour. Phila. Acad. i, 1817.)

Schiodte et Meinert. Monograph of the Cymothoidæ. (Naturhist. Tidssk. 1879–.) (Lat.)

Smith, S. I. New or little-known American Cancroid Crustacea. (Proc. Bost. Socy. xii. 1869.)

Smith, S. I. Crustacea collected by C. F. Hartt in Brazil. (Trans. Conn. Acad. i, 1869.)

Smith, S. I. **Notes on N. A. Crustacea: I,** Ocypodoidea. (Trans. Conn. Acad. i. 1870.)

Smith, S. I. **Crustacea in Report U. S. Fish Commission.** 1871–2. 8vo. 1873.

Smith, S. I. **Crustacea of fresh waters of U. S.** (Rep. U. S. Fish Com., 1872–3. (1874.)

Smith, S. I. **Stalk-eyed Crustacea of Atlantic Coasts of North America.** (Trans. Conn. Acad. v, 1879.)

Stimpson, W. **Crustacea and Echinodermata** of Pacific Coast of N. A. (Jour. Bost. Socy. vi, 1857).

Stimpson, W. **Crustacea of the North Pacific Exploring Expedition.** (Proc. Phila. Acad., 1858–60). (Lat.)

Stimpson, W. **Notes on North American Crustacea.** (Annals N. Y. Lyceum vii and x. 1859–71).

Stimpson, W. **Crustacea dredged in the Gulf Stream.** (Bulletin Mus. Comp. Zool. ii, 1871).

Stuxberg, A. North American Oniscidæ. (K. Vet. Akad. Forh. 1875. Sw. and Lat.)

Verrill, A. E. Phyllopoda of family Branchipidæ. (Proc. Am. Assoc. Adv. Sci. 1869.)

Weismann, A. Contributions to a Natural History of the Daphnoidea. (Zeitsch. Wiss. Zool. 1876-8. (Ger.)

White, A. List of Crustacea in British Museum. 12 mo. 1847.

Wood, Mason J. Indian and Malayan Thelphusidæ. (Jour. Asiatic Socy. Bengal xl. 1871.)

De Haan W. Fauna Japonica, Crustacea, folio, 70 pls. Leyden, 1835-50. (Lat.)

MOLLUSCA.

Adams and Reeve. Mollusca of the voyage of the Samarang. 4to, 24 pls. London, 1848.

Adams, C. B. Catalogue of shells collected at Panama. 8vo. N. Y., 1852.

Adams, C. B. Contributions to Conchology. 8vo. N. Y., 1849-52.

Adams, H and A. Genera of recent Mollusca. 2 vols. 8vo. atlas, 144 pls. London, 1853-8.

Alder and Handcock. Monograph British Nudibranch Mollusca. (Ray Socy.) 2 vols. fol. 83 pls. London, 1845-55.

Bergh, R. Malacological Researches. Scientific Results of Semper's Travels in the Philippines. 4to, 58 pls. Weisbaden, 1870-77. (Ger.)

Bergh, R. On the Nudibranchiate Gasteropod Mollusca of the North Pacific Ocean. (Proc. Phila. Acad. 1879.) See numerous other papers on Nudibranchiata, etc. Copenhagen et Vienna, 1853.

Binney, A. Terrestrial air-breathing Mollusca of the U. S. Edited by A. A. Gould. 4 vols., 8vo, 110 pls. Boston, 1851-59. (See also Jour. Bost. Socy. Vol. ii, et seq.)

Binney, W. G. Bibliography of N. A. Conchology.
(Smithsonian Misc. Coll. V.) 2 pts 8vo. Washington, 1863-64

Binney, W. G. Catalogue Terrestrial air-breathing Molluscs of North America. (Bulletin M. C. Z. iii.) 1874.

Binney, W. G. Notes on American land shells. 8vo. Burlington, N. J., 1874-75.

Binney, Bland, and Tryon. Land and Fresh Water Shells of North America. (Smithsonian Misc. Coll.) 4 vols. 8vo, 1865-74.

Blainville, de, D. Manual of Malacology and Conchology. 8vo. 109 pls. Paris, 1825. (Fr.)

Blanford, W. T. and H. F. Contributions to Indian Malacology. 12 nos. 8vo. Calcutta, 1860-70.

Bourguinat, J. R. Malacology of Algiers. 2 vols. 4to, 60 pls. Paris, 1863-64. (Fr.)

Bourguinat, J. R. Molluscs, new, doubtful or little known. 12 parts, 8vo, 49 pls. Paris, 1863-70. (Fr.)

Brot, A. Catalogue of the genera, species and varieties of Melanidæ. 8vo. New York, 1868.

Brown, T. Illustrations of the recent Conchology of Great Britain. 2 edit., 4to. London, 1844.

Calkins, W. W. Marine Shells of Florida. (Proc. Dav. Acad., 1878.

Carpenter, P. P. Molluscs of Western North America. (Smithsonian Misc. Coll., vol. x. Reprints of many of his papers on the subject.) 1872.

Carpenter, Lea, Stimpson, etc. Check List North American shells. (S. I. Misc. Coll., ii, 1860.)

Catlow and Reeve. Conchologist's Nomenclator, a catalogue of all the recent species of shells. 8vo. London, 1845.

Chenu, J. C. Bibliothèque Conchyliologie. [Papers by Say, Conrad, Rafinesque, Leach, etc.] 3 vols., 8vo, 91 pls. Paris 1845-6. (Fr.)

Chenu, J. C. Manual of Conchology. 2 vols., 8vo, 5,000 fig. Paris, 1860-62. (Fr.)

Conrad, T. A. Catalogue and Synonymy of Genera and Species of Solenidæ, Mactridæ and Anatinidæ. 8vo. Philadelphia, 1868.

Conrad, T. A. Synopsis of American Naiades. 1 vol., 8vo. Philadelphia, 1834.

Cox, J. Monograph Australian Land Shells. 4to. Sydney, 1868.

Dall, W. H. Limpets and Chitons of Alaska, etc. Proc. Nat. Mus. i, 1879.

Dall, W. H. The Genus Pompholyx and its allies with a revision of the Limæidæ. (Am. N. Y. Lyc. ix, 1870.)

Dall, W. H. Brachiopoda obtained by U. S. Coast survey with a revision of the Craniidæ and Discinidæ. (Bull. M. C. Z.) 1871.

Dekay, J. E. N. Y. Fauna: Pt. V, Mollusca. 4to. Albany, 1843.

Deshayes, G. B. Mollusca in Regne animal. 4to, 222 pls. Paris, 1850. (Fr.)

Deshayes, G. B. Catalogue of Bivalve Shells in British Museum. 8vo. London, 1853-54.

Donovan, E. Natural History of British shells. 5 vols., 8vo, many plates. London. 1799-1804.

D'Orbigny. Molluscs of de la Sagras Cuba. 2 vols., 8vo, with folio Atlas. Paris, 1853-55. (Fr.)

Draparnaud, J. Natural History of Land and fluviatile Mollusca of France. 2 vols., 4to. Paris. 1805-31. (Fr.)

DuClos, P. L. Monograph of genus Oliva. fol., 35 pls. Paris, 1835. (Fr.)

DuClos, P. L. Natural History of Oliva and Columbella. fol., 47 pls. Paris, 1835-40.

Eydoux et Souleyet. Mollusca of Voyage of La Bonite. Paris, 1836-7. (Fr.)

Ferrusac et Deshayes. Natural History of Molluscs living and fossil. 42 pts., folio, 247 pls. Paris, 1820-51. (Fr.)

Fischer et Crosse. Studies of the terrestrial and Fluviatile Molluscs of Mexico and Central America. Fol., many plates Paris, 1870 . (Fr.)

Forbes and Hanley. History of British Mollusca. 4 vols., 8vo, 202 pls. London, 1853.

Gould, A. A. Report on the Invertebrata of Massachusetts. 8vo. Cambridge, 1841. **Second Edition by W. G. Binney.** Boston, 1869.

Gould, A. A. Mollusca of U. S. Exploring Expedition. 4to, with atlas of Many plates. Philadelphia, 1852.

Gould, A. A. Otla Conchologica. [Comprising shorter papers from 1839-62]. 8vo. Boston, 1862.

Gray, J. E. Catalogues of Mollusca in British Museum. 9 parts. London, 1849-57.

Haldemann, S. S. Monograph of fresh-water univalve Mollusca of U. S. 8vo, 35 pls. Philadelphia, 1840-44.

Hanley, S. Ipsa Linnæi Conchylia. 8vo. London, 1855.

Hanley, S. Conchological Miscellany. 4to, 40 pls. London, 1858.

Hanley and Theobald. Conchologia Indica. 4to, 160 Pls. London, 1870-76.

Hidalgo, J. G. Marine Mollusca of Spain, Portugal and the Balearic Isles. 8vo, many plates. Madrid, 1870-. (Sp.)

Jay, J. C. Catalogue of Recent shells in his cabinet, 4th edit., 4to. 1852.

Jeffrey, J. G. British Conchology. 5 vols., 8vo, 147 pls. London, 1862-69.

Kiener, L. C. Spécies générale et Iconographie des Coquilles vivants. 149 pts. pub. 861 pls., 8vo. Paris, 1839-74. (Fr.)

Kobelt. Illustrated shell book. 4to, many plates. Nuremburg, 1876. (Ger.)

Küster, H. C. The Family Volutacea. 3 pts., 4to, 56 pls. Nuremburg, 1841-49. (Ger.)

Küster and Weinkauff. Monograph of Conidæ. 4to, 72 pls. Nuremburg, 1875. (Ger.)

Lamarck. Historie Naturelle des Animaux sans Vertebres. 2nd edit. by DeShayes. 11 vols., 8vo. Paris, 1835-45.

Lea, Isaac. Observations on the genus Unio. 13 vols., 4to, 250 pls. Philadelphia, 1834-74. [Several parts in Trans. Am. Phil. Socy., Jour. Acad. Nat. Sci. and Am. Jour. Sci.]

Lea, Isaac. Synopsis of the Family Naiades. 1st edit., 8vo, 1836. 4th Edit. 4to. Philadelphia. 1870. (See also over 150 other papers on Land & Fluviatile Mollusca in Trans. Am. Phil. Socy. iii, 1830-x, 1853; Proc. Am. Phil. Socy., i, 1838-v, 1854. Proc. Phila.; Acad. 1841-73. Jour. Phil. Acad. II, ii, 1854-viii, 1874; Proc. Zoöl. Soc. Lond., 1850.)

Lischke, C. E. Japanese Marine Shells. 3 pts. 4to. Cassel, 1869-75. (Ger.)

Martini et Chemnitz. Systematische Conchylien Cabinet. 270 pts. pub. 4to. 1400 pls. Nuremburg, 1840-78. (Ger.)

Moquin Tandon, A. Terrestrial and Fluviatile Mollusca of France. 2 vols. 8vo. 50 pls. Paris, 1855. (Fr.)

Mörch, O. A. S. Descriptive Catalogue West Indian Scalidæ. Jour. Phila. Acad. viii, 1876.

Morse, E. S. Observations on the Terrestrial Pulmonifera of Maine. (Jour. Portland Socy. vol i, 1864.

Morse, E. S. Land Shells of New England. (Amer. Naturalist i, 1867.)

Neville, G. Hand List of Mollusca in the Indian Museum. 8vo. Calcutta, 1878.

BIBLIOGRAPHY.

Pfeiffer. Monographia Heliceorum vivent. 8vols. 8vo. Leipzig, 1848-77.

Pfeiffer. Novitates Conchologicæ. 67 pts. Cassel, 1855-77. (Ger.)

Philippi, R. A. Figures and descriptions of new or little known shells. 3 vols. 4to. 144 pls. Cassel, 1845-51. (Ger.)

Prime, T. Monograph of North American Corbiculadæ (Smithsonian Misc. Coll. vii) 1865.

Prime, T. Catalogue and synonymy of the recent genera and species of Corbiculadæ. 8vo. Philadelphia, 1868.

Rafinesque, C. S. Complete writings on Conchology. Edited by Binney and Tryon. 8vo. Philadelphia, 1864.

Redfield, J. H. Catalogue and synonymy of the species of Marginellidæ. (Am. Jour. Conch., 1870.)

Reeve, L. Conchologia Iconica, Monographs of the genera of Shells, continued by G. B. Sowerby. 20 vols. 4to. 2727pls. 20,000 figures. London, 1843-78.

Römer, E. Monographs of Dosiniæ, Venides, Cytherea, Donax, Cardium and Tellina. 4to 264 pls. Cassel, 1862-72. (Ger.)

Sander Rang, A. Natural History of the Aplysieus. 4to. 25 pls. Paris, 1828. (Fr.)

Sander Rang and Souleyet. Natural History of the Pteropod Mollusca. 4to. Paris, 1852. (Fr.)

Sars, G. O.. Mollusca Regionis Arctica Norvegiæ. 8vo. Christiania, 1878

Say, T. Description of the Shells of North America. 8vo. 68 pls. Philadelphia, 1830-34.

Smith and Prime. Mollusca of Long Island. (Ann. N. Y. Lyc. ix, 1870.

Sowerby, G. B. Conchological Manual. 4th edit. 8vo. London, 1852.

Sowerby, G. B. Conchological Illustrations. 200 pts. 200 pls. 8vo. London, 1841-45.

Sowerby, G B. Thesaurus Conchyliorum. 32 pts. 4to. 350 pls. London, 1842-66.

Sowerby, J. Genera of Recent and Fossil Shells. 42 nos. 8vo. 266 pls. London, 1820-25.

Stimpson, W. Shells of New England. 8vo. Boston, 1851.

Stimpson. W. Monograph Hydrobinæ. (Smithsonian Misc. Coll. vii, 1865.

Swainson, W. Exotic Conchology. 2 edit. by Hanley. 4to. 48 pls. London, 1841.

Tryon, G. W. American Marine Conchology. 8vo. 44 pls. Philadelphia, 1873.

Tryon, G. W. Monograph of the Pholadacea. 8vo. Philadelphia, 1862.

Tryon, G. W. Catalogues and Synonymy of Pholadacea, Saxicavidæ, Myidæ, Corbulidæ, Tellenidæ, Galeomiuidæ, Astartidæ, Solemyidæ, Lucinidæ, Chemidæ and Chametrachidæ. 8vo. Phila., 1862-72. (Am. Jour. Conch. and Proc. Phil. Acad.)

Tryon, G. W. North American Strepomatidæ (Melanians). S. I. Misc. Coll. xii, 1873.

Tryon, G. W. Manual of Conchology. 8vo. Many *poor* plates. Phila., 1879 *et seq*.

Verrill, A. E. Descriptions of shells from Gulf of California. Am. Jour. Sci., 1870

Verrill, A. E. Synopsis of Cephalopoda of the northeast coast of America. (Am. Jour Sci., 1880.)

Verrill, A. E. Cephalopods of the northeast coast of America. Pt. I. Trans. Conn Acad. v, 1879.

Verrill, A. E. Mollusca of Vineyard Sound. In Rep. U. S. Fish Comm., 1871-72. (1873.) See also numerous other papers in Am. Jour. Sci , etc.

Wood, W. Index Testaceologicus. 2 edit. by Hanley. 8vo. 2780 figs. London, 1856.

Woodward, S. P. Manual of the Mollusca. 8vo.

Lond., 1851-56. 3 edit. London, 1875. (The first edition contains the Tunicata which are omitted from the others.)

Yarrow, H. C. Mollusca in Rep. Wheeler's Survey, v, 1875.

MOLLUSCOIDA, TUNICATA, BRACHIOPODA, and POLYZOA.

(See also general works on Mollusca.)

Allman, G. J. Monograph of Fresh Water Polyzoa British and Foreign. (Ray Soc'y.) fol. 11 pls. London, 1856.

Busk, F. Catalogue of Marine Polyzoa in British Museum. 3 pts. (1 and 2, 12mo; 3, 8vo.) 162 pls. London, 1852-75.

Dall, W. H. Catalogue of recent species of Brachiopoda. (Proc. Phila. Acad. 1873.)

Dall, W. H. Index to names which have been applied to the subdivisions of the Brachiopoda. (Bulletin U. S. National Museum. No. 8, 1877.

Davidson, T. Classification of the Brachiopoda. London, 1853. (See also Proc. Zool. Socy., 1871, etc.)

Dumortier et Van Beneden. Natural History of the Fluviatile Bryozoa. 4to, 6 pls. Brussels, 1850. (Fr.)

Heller, C. Bryozoa of the Adriatic. 8vo. Vienna, 1867. (Ger.)

Heller, C. Researches on the Tunicata of the Adriatic. (Denksch. K. Akad. Wiss., Wien, 1874-77) 4to, 19 pls.

Hyatt, A. Observations on Fresh Water Polyzoa. (Proc. Essex Inst. iv-v.) 1866-68.

Lacaze-Duthiers, H. Natural History of Mediterranean Brachiopoda. 8vo. Paris, 1868.

Milne-Edwards, H. Compound Ascidia of la Manche. (Arch. du Mus., 1872) 4to, 18 pls. 1872-3. (English,)

Smitt, F. A. Floridan Bryozoa. (Trans. Roy. Sw. Acad.) 4to, 18 pls. 1872-3. (English.)

Smitt, F. A. Scandinavian Marine Polyzoa. 5 pts., 4to, 24 pls. Stockholm, 1865-72. (Sw.)

Vogt, C. Siphonophora and Tunicata of Nice. 2 vols., 4to, 27 pls. Geneva, 1854. (Fr.)

WORMS.

Audouin et Milne-Edwards. Researches in Natural History of France. 2 vols., 8vo, 25 pls. Paris, 1832-34. (Fr.)

Audouin et Milne-Edwards. Classification of Annelides. 8vo, 18 pls. Paris, 1834. (Fr.)

Baird, W. Catalogue of Entozoa in British Museum. 8vo. London, 1853.

Van Beneden, C. J. Memoire sur les Vers Intestinaux. 4to, 27 pls. Paris, 1861. (Fr.)

Van Beneden, C. J. Turbellarians of Belgium. 4to. Brussels, 1861. (Fr.)

Van Beneden et Hesse. Marine Bdellidæ and Trematoda. 4to, 18 pls. Brussels, 1863-65. (Fr.)

Claparede, E. Annelida Chætopoda of Naples. 4to, 46 pls. Geneva, 1867-71. (Fr.)

Cobbold, T. S. Entozoa. 8vo. London, 1879.

Diesing, C. M. Systema Helminthum. 2 vols., 8vo. Vienna, 1850-51. (Numerous other papers in Publications of Vienna Societies, 1836.

D'udekem. New classification of the setigerous Annelid Abranchiata. (Mem. Acad. Roy. Belg. 1859.) (See also Bulletin Acad. Belg. 1855.)

Dujardin, F. Natural History of the Helminthes. 8vo. Paris, 1845. (Fr.)

Ehlers, E. Die Borstenwürmer, Annelida Chætopoda. 2 pts., 4to, 23 pls Leipzig, 1864-68. (Ger.)

Hubrecht, A. W. Genera of European Nemerteans critically revised. (Note xliv of Leyden Museum.) 1878. (Fr.)

Jensen, O. S. Turbellaria ad Litoria Norwegiæ occidentale. Bergen, 1878.

Keferstein. **Contributions to Systematic Knowledge of Sipunculidæ.** (Zeit. Wiss. Zool. xvi.) 1865. (Ger.)

Keferstein. On American Sipunculids. (Zeit. Wiss. Zool. xvii.) 1866. (Ger.)

Kramer. The Family Bdellidæ. (Arch. für Naturgesch. xlii.) (Ger.)

Leidy, Jos. Flora and Fauna in Living Animals. (Smithsonian Cont. v.) 1853.

Leidy, Jos. Helminthological contributions. (Proc. Phil. Acad., 1850.)

Leidy, Jos. Synopsis of Entozoa. (Proc. Phil. Acad., 1856.)

Leuckart, R. Human Parasites. 2 vols., 8vo. Leipzig, 1863–68. (Ger.)

Malmgren, A. J. Marine Arctic Annulata. 8vo, 20 pls. Stockholm, 1865–66.

Malmgren, A. J. Polychætous Annelida of Spitzbergen, Greenland, etc. 8vo, 14 pls. Stockholm, 1867. (Lat.)

McIntosh, W. C. **Monograph of British Annelids: I, Nemerteans.** (Ray Socy.) fol., 23 pls. London, 1873–74.

McIntosh, W. C. On British Annelida. Pt. I, 4to. (London) 1876.

Moquin Tandon. Monograph of the Hirudinidæ. 8vo, 14 pls., 2 edit. Paris, 1846. (Fr.)

Mörch, O. A. L. Critical Revision of the Serpulidæ. 8vo. Copenhagen, 1863. (Lat.)

Moseley, H. N. Notes on Land Planarians with a list of known species. (Micros. Jour., xvii. 1877.)

Oersted, A. S. Gronlands Annulata dorsibranchiata. 4to, 8 pls. Copenhagen, 1843. (Dan.)

Perrier, E. Researches on the history of the terrestrial Lumbricidæ. 4to, 5 pls. (Paris), 1872. **(Fr.)**

Pourtales, L. F. de. American Sipunculids. (Proc. Am. Assoc., 1851.)

Quatrefages, A. de. History of the Annelida and Gephyrea. 3 vols., 8vo, 20 pls. Paris, 1865. (Fr.)

Rudolphi, C. Entozoörum sur Verinium Intestinum Historia Naturalia. 2 vols., 8vo, 12 pls. Amsterdam, 1808-10. (Lat.)

Rudolphi, C. Entozoörum synopsis. 8vo. Berlin.

Savigny, J. C. Systeme des Annelides. Fol. Paris, 1820. (Fr.)

Schmarda, L. K. Neue wirbellose Thiere. (Turbellaria, Rotatoria et Annelida). 2 pts., folio, 37 pls. Leipzig, 1859-61. (Ger.)

Schultze, M. Natural History of the Turbellaria. 4to, 7 pls. Griefswald, 1851. (Ger.)

Stimpson, W. Prodromus descriptionis animali, etc. Pts. 1 & 2. Turbellaria. Proc. Phil. Acad., 1857. (Lat.)

Verrill, A. E. Synopsis of North American fresh water leeches. (Rep. U. S. Fish Commission. 1872-3. 1874.)

Verrill, A. E. Leeches in Report Wheeler's Surv., v. 1875.

Villott, A. Monograph of Gordiadæ. Arch. Zool. Exp. et gen., iii. 1874. (Fr.)

Webster, H. E. Annelida Chætopoda of Virginia Coast. (Trans. Albany Inst., ix. 1879).

Williams, T. Report on British Annelida. 8vo, 10 pls. (London), 1852.

ECHINODERMATA AND CŒLENTERATA.

Agassiz, A. Illustrated Catalogue of North American Acalephæ. (Ill. Cat. Mus. Comp. Zool.) 4to. Cambridge, 1865.

Agassiz, A. List of Echinoderms sent by the Museum of Comp. Zoology to various Institutions. (Bulletin M. C. Z. i, 1863.)

Agassiz, A. Revision of the Echini. (Memoirs M. C. Z.) 4 pts. 94 pls. 1872-74.

Agassiz, A. and E. C. Seaside studies in Natural History. 8vo. Boston, 1871.

Agassiz, Lyman and Pourtales. Contributions to the fauna of the Gulf Stream at great depths. (B. M. C. Z. i, 1869.)

Agassiz and Pourtales. Echini, Crinoids and Corals of the Hassler Expedition. (Ill. Cat. M. C. Z. viii, 1874.)

Agassiz and Pourtales. Monograph of Corals of Florida. (Ill. Cat. M. C. Z., 1871.)

Agassiz, L. Monographie d'Echinodermes. 4 vols. 4to. 63 fol. pls. Neuchatel, 1838-42. (Fr.)

Agassiz, L. Contributions to Natural History of the Acalephæ of the U. S. (Mem. Am. Acad., 1849.)

Agassiz, L. Contributions to the Natural History of the U. S. 4 vols. 4to. many plates, Boston, 1857-62.

Allman, G. J. Monograph of the Gymnoblastic or Tubularian Hydioids. 2 vols. fol. (Ray Soc.) 23 pls. London, 1871-72.

Van Beneden, P. J. Polyps of Belgium. 4to. 19 pls. Brussels, 1866.

Clark, H. E. Prodrome of Lucernaria. (Jour. Bost. Socy. 1863.)

Clarke, S. F. New Hydroids. (Am. Jour. Sci. iii, 1876.)

Clarke, S. F. Alaskan Hydroids. (Proc. Phila. Acad. 1877.)

Dana, J. D. Zoophytes U. S. Exploring Expedition. 4to, with folio atlas, 61 pls. New Haven, 1846-59.

Dana, J. D. Structure and Classification of Zoöphytes. 4to. Phila., 1846.

Dana, J. D. Coral and Coral Islands. 8vo. N. Y., 1872.

Danielssen and Koren. Results of the Norse Northern Expedition. 3 pts. pub. 8vo. 1877-79.

Düben et Koren. Review of Scandinavian Echinodermata. 8vo. Stockholm, 1845.

Duchassaing et Michelotti. Corals of the Antilles. 2 pts. 4to. 21 pls. Turin, 1860-64. (Fr.)

Esper, E. J. Die Pflanzenthiere. (Zoophyta.) 3 pts. and suppl. 4to. 443 pls. Nurenburg, 1791-1830. (Ger.)

Forbes, E. History of British Starfishes and other Echinodermata. 8vo. London, 1841.

Forbes, E. Monograph of British Naked-eyed Medusæ. (Ray Soc.) fol. 13 pls. London, 1846.

Gegenbaur, C. Contributions to a better knowledge of the Siphonophora. 4to. 10 pls. Leipzig, 1854-60.

Gosse, P. H. Actinologia Brittanica. (British Sea Anemones and Corals.) 8vo. London, 1860.

Gray, J. E. Catalogues of Echinida, Starfishes, Sea Pens and Stony Corals in British Museum. Zoöl. 8vo. 1855-70.

Gray, J. E. Synopsis of the Star Fish in British Museum. 4to. 16 pls. London, 1867.

Greene, J. R. Manual of the Cœlenterata. 8vo. London, 1869.

Haeckel, E. Contributions to Nat. Hist. of Hydromedusæ. 8vo. Leipzig, 1865.

Haeckel, E. System der Medusen. 4to, many plates. 1879. (Ger.)

Hincks, T. Natural History of British Hydroid Zoophytes. 2 vols. 8vo. 67 pls. London, 1868.

Hulton, F. W. Catalogue New Zealand Echinodermata. 8vo. Wellington, 1872.

Huxley, T. H. Oceanic Hydrozoa. (Ray Soc.) 4to. 12 pls. London, 1859.

Johnston, G. History of British Zoophytes. 8vo. 44 pls. Edinburg, 1838; 2 edit. 74 pls. London, 1847.

Kolliker, A. Siphonophora of Messina. fol. 12 pl. Leipzig, 1853. (Ger.)

Kolliker, A. Descriptions, Anatomical and Systematic, of the Alcionaria. (Pt. I, 4to, 24 pls. Frankfort, 1872. (**Ger.**)

Lamouroux, J. G. History of the flexible Coralline polypidoms. 8vo, 19 pls. Caen, 1816. (Fr.) English Edition, London, 1824

Leidy, J. Marine Invertebrate Fauna of Rhode Island and New Jersey. (Jour. Phil. Acad., iii, 1855).

Lockington. List of Echinoidea in Collection of California Academy. (Proc. Cal. Acad., vii, 1875.)

Lütken, C. Echinoderms of Central America. 8vo. Copenhagen, 1858.

Lütken, C. Additamenta ad historiam Ophiuridarum. 4to, 7 pls. Copenhagen, 1859-69. (Lat.)

Lyman, T. Ophiuridæ and Astrophytidæ. (Ill. Cat. Mus. Comp. Zool., No. 1, 1864. Suppl. No. 6, 1871.)

Lyman, T.' Ophiuridæ and Astrophytidæ, new and old. (Bulletin M. C. Z., iii, 1874).

Lyman, T. Ophiuridæ and Astrophytidæ of the Hassler Expedition. Memoirs M. C. Z, 1874.

Macready, J. Proc. Elliot Socy., i. 1856-60.

Mertens, H. Observations and Researches on Beroid Acalephæ. 4to, 13 pls. St. Petersburg, 1833. (Ger.)

Milne-Edwards, H. Researches on the Polyps. 8vo, 28 pls. Paris, 1838. (Fr.)

Milne-Edwards, H. Natural History of the Polyps proper. 8vo, 31 pls. Paris, 1857-60. (Fr.)

Müller und Troschel. System der Asteriden. 4to, 12 pls. Brunswick, 1842. (Ger.)

Pervier, E. Revision of Stellierdes. (Arch. Zool. Exp. et Gen., iv, 1875.) (Fr.)

Pourtales, L. F. de. Deep Sea Corals. (Illust. Catalogue M. C. Z., iv, 1871).

Pourtales, L. F. de. Corals and Antipathea of the Caribbean Sea. (Bulletin Mus. Comp. Zool., vi, 1880.)

Rathbun, R. Echinoid Fauna of Brazil. (Am. Jour. Sci., 1878.)

Rathbun, R. List of Brazilian Echinoderms. (Trans. Conn. Acad., v, 1878).

Richiardi, S. Monograph of Pennatulidæ. 8vo, 14 pls. Bologna, 1869. (Ital.)

Sars, M. Memoir on living Crinoids. 4to, 6 pls. Christiana, 1868. (Fr.)

Sars, M. Review of Norse Echinodermata. 8vo. Christiana, 1861.

Savigny, J. C Iconographie des Echinodermes, Polypes et Zoophytes de l'Egypt, fol. 28 pls. Paris, 1810.

Selenka, E. Anatomy and Revision of the Holothurians. (Zeitsch. Wiss. Zool. xvii.) 1867. 4 pls. (Ger.)

Semper, C. Scientific Results of Travels in Philippene Archipelago. Holothurians. 4to, 57 pls. Leipzig, 1867-8. (The most valuable work on the subject ever published.)

Trywell, G. Manual of British Sea Anemones. 8vo, 7 pls. London, 1856.

Verrill, A. E. Revision of Polypes, east coast of America. (Memoirs Bost. Socy. i, 1863.)

Verrill, A. E. List of Polypes and Corals sent by Museum Comp. Zool. (Bulletin M. C. Z.) 1, 1864.

Verrill, A. E. Synopsis of Polypes and Corals of No. Pacific Exploring Expedition. (Bulletin Essex Inst. iv-vi, 1866-69.)

Verrill, A. E. Notes on Radiata. (Trans. Conn. Acad. i, 1868-71.)

Verrill, A. E. New and imperfectly known Echinoderms and Corals. (Proc. Bost. Socy. xii. 1869.)

Vogt, C. Siphonophora and Tunicata of Nice. 2 vols., 4to, 27 pls. Geneva, 1854. (Fr.)

SPONGES AND PROTOZOA.

Bowerbank, J. Monograph of British Spongiadæ.
3 vols. 8vo, 129 pls. (Ray Soc.) London, 1864-74.
Bowerbank, J. Monograph of Siliceo-fibrous sponges. 8vo. London, 1869-76.
Butschli, O. Contributions to a knowledge of the Flagellata. (Zeit. Wiss. Zoöl. xxviii, 1878.) (Ger.)
Carpenter, W. B. Researches on Foraminifera. 4to. 22 pls. London, 1856-61.
Carpenter, Parker and Jones. Introduction to study of Foraminifera. fol. 22 pls. (Ray Soc.) London, 1862.
Claparede et Lachmann. Studies of Infusoria and Rhizopoda. 2 vols. 4to, 37 pls. Geneva, 1858-61 (Fr.)
D'Orbigny, A. Foraminifera in de la Sagra's Cuba. Paris, 1839. (Fr.)
Duchassaing et Michelotti. Sponges of the Caribbean and Antilles. 4to. 25 pls. Harlem, 1864. (Fr.)
Dujardin, F. Natural History of the Infusoria. (Suites à Buffon). 8vo. Paris, 1841. (Fr.)
Ehrenberg, C. G. Organization, Classification and Geographical Distribution of the Infusoria. 4to. 24 pls. Berlin, 1830-36. (Ger.)
Ehrenberg, C. G. Die Infusionthierchen. fol. 64 pls. Leipzig, 1838. (Ger.)
Greene, J R. Manual of the Protozoa. 8vo. London, 1871.
Haeckel, E. Monograph of the Radiolaria. Folio with atlas, 35 pls. Berlin, 1862. (Ger.)
Haeckel, E. Monograph of Monera. 8vo. Jena, 1868. (Ger.)
Haeckel, E. Monograph of Calcispongiæ. 3 vols. 4to. 60 pls. Berlin, 1872. (Ger.)
Hertwig R. Studies on Rhizopoda. (Jena Zeitsch., 1877. Ger.)

Hyatt, A. Revision of North American Poriferæ.
(Mem. Bost. Soc. ii, 1874.)

Johnston, G. History of British Sponges. 8vo. Edinburg. 1842.

Leidy, J. Flora and Fauna in living Animals. (Smithsonian Contributions v, 1853.)

Leidy, J. Fresh Water Rhizopoda of North America. (Hayden's Survey.) 4to, 48 pls. 1880.

Parker and Jones. Nomenclature of the Foraminifera. 10 pts., 8vo. (London), 1859-72.

Parker and Jones. On some Foraminifera from the North Atlantic. 4to, 8 pls. (London, 1865.)

Pritchard, A. History of the Infusoria. 8vo. London, 1842. **2 edit.** 8vo, 13 pls. London, 1849.

Quennerstedt, A. Contributions to the Swedish Infusoria fauna. 4to (Lund), 1866-70. (Swed.)

Stein, Fr. Die Organismus der Infusionsthiere. 3 vols., 4to, many pls. Leipzig, 1859-77. (Ger.)

Williamson. Monograph Recent Foraminifera. (Ray Socy.) fol. London, 1857.

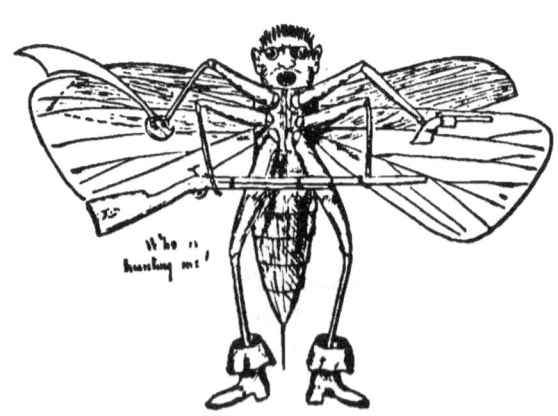

INDEX.

	PAGE		PAGE
Aberration	82	Breeding cage	31
Acetic acid	137	Breeding larvæ	31
Adjustment of microscope	85	Bullock's arsenical powder	141
Adjustment for cover-glass	98		
Alcohol	41	Bulls eye	90
Alcoholic specimens	3	Butterfly nets	19
Ammonia carmine	135	Butterfly triangles	25
Angle of aperture	98		
Arsenical powder	141	Cabinets for bottles	74
Argentic nitrate	137	Camera lucida	91, 99
Arsenical soap	140	Card catalogues	51
Artificial serum	138	Care of microscope	96
Auric chloride	37	Caring for collections	40
		Carmine	135
Batrachia	12	Cases	63, 68
Bausch and Lomb microscope	95	Cases for microscopic slides	75
Beam trawl	37	Cases for small bottles	73
Beck's microscopes	95	Catalogues	50
Beating for insects	20	Cement	142
Becœur's arsenical soap	140	Chairs	106
Bibliography	149	Chloride of gold	137
Bichromate of potash	38	Chromic acid	125
Bird lime	2, 3	Clams, dissecting	118
Birds	6	Coarse adjustment of microscope	85
Blackening brass	143		
Blackening wax	57	Coddington lens	82
Black ink	143	Collecting and preserving, works on	1
Blow gun	2		
Borax carmine	135	Collecting batrachia	12
Boring glass	57	Collecting equipment	31
Bottom collecting	35	Collecting insects	18
Bottles	55	Collecting net	18
Brackets for shelves	70	Collecting reptiles	12
Brass, to blacken	1, 3	Collecting umbrella	20

INDEX.

	PAGE
Collecting vertebrates	2
College museums	64
Colors; conventional	59
Compound microscope	84
Compressórium	101
Condenser	90
Conventional colors	59, 110
Corrosive sublimate	141
Craig microscope	92
Crustacea	118
Cyanide bottle	20
Cyanide of potassium	20
Destroying pigment	125
Diaphragm	90
Dissecting	115
Dissecting insects, etc.	116
Dissecting microscopes	84
Dissecting tank	115
Dissections preserved	56
Dissolving paraffine	129
Drawing	109
Drawtube	85
Dredging	35
Eggs	11
Egg drill	11
Eggs exhibiting	52
Elder pith	128
Electrical cement	142
Eosin	137
Equipment, collecting	34
Exhibition cases	69
Exhibiting birds' eggs	52
Fine adjustment of microscope	85
Fishes	13
Focal length of objectives	87
Focussing the microscope	97
Formulæ	135
Freezing microtome	129
Freezing tissues	129
Frey's fuschine	136
Frogs	119
Fruit jars	41, 57
Fuschine	136

	PAGE
Gasteropoda	118
Gelatine injections	121
Generic names	46
Glass, to bore	57
Glass stages	89
Glycerine and gum	139
Glycerine jelly	121, 138
Goadby's solutions	140
Gold chloride	137
Grafting wax	54, 143
Gum	48
Gum arabic	127
Gumming insects	23
Hæmatoxylin	135
Hardening tissues	123
Hartnack microscopes	94
Hartnack objectives	89
Heliotype	110
High angle lenses	98
"Homœopathic" collections	73
Homœopathic vials	56
Horizontal cases	71
Huygenian oculars	86
Ichneumon parasites	33
Illustrations	109
Imbedding	125, 127
Imbedding tray	126
Immersion lenses	87, 97
India ink	46
Inflating larvæ	27
Inflating oven	28
Injecting	120
Injecting media	121
Ink	46, 143
Ink for labels	46
Insects	18
Insect cases	72
Insect forceps	24
Insect labels	49
Insect localities	22
Insect net	23
Insect pins	23
Insect poison	142
Instruments for laboratory	107

INDEX.

	PAGE
Iodized serum	138
Jars for storage	57
Jelly fish	41
Killing insects	20
Killing spiders	30
Killing marine forms	41
Kleinenberg's hæmatoxylin	136
Labels	45
Label holders	48
Labels, large	59
Labelling birds	9
Labelling bottles	49
Labelling fossils and minerals	50
Labelling insects	49
Laboratories	67, 105
Laboratory necessaries	107
Laboratory tables	105
Laboratory work	115
Lamellibranchs	118
Land shells	42
Large labels	59
Larvæ, breeding	81
Larvæ, inflating	27
Laurent's arsenical soap	141
Leconte's insect poison	142
Lenses	81
Lens holder	83
Lifting sections	129
Lobsters	118
Locks	69
Logwood (see Hæmatoxylin)	135
Macerating skeletons	16
Macerating skulls	17
Mammals	4
Marine collecting	34
Medusæ	41
Mending insects	26
Microscope	81
Microscopic slide cases	75
Microtomes	127, 130
Moistenring insects	25
Moleschott's acetic acid	137
Mounting fishes	14

	PAGE
Mounting shells	53
Mounting skeletons	17
Mounting specimens	45
Mounting spiders	30
Mucilage	48, 142
Müllers fluid	124, 138
Museum plans	64
Myriapoda	33
Nachet objectives	89
Natural skeletons	17
Nests and eggs	11
Neutral salt solution	138
Nitric acid	125
Nitrate of silver	137
Note books	109
Novelty microscope	92
Objectives	84
Oculars	84
Old alcohol	143
Oniscidæ	34
Osmic acid	124, 137
Ox gall for mending insects	26
Packing butterflies	25
Packing insects	26
Packing jars	57
Painting tablets	53
Paper trays	126
Paraffine	125
Peron's luting	143
Perosmic acid	137
Photo-illustrations	111
Picrocarmine	135
Pigment	125
Pill bugs	34
Pinning forceps	24
Pinning insects	23
Plan for museum	64
Poison bottle	20
Poisoning insects	20
Polariscope	91
Polyzoa	42
Potassic bichromate	138
Protozoa	117
Printed labels	46
Pumping	58

	PAGE
Quinine bottles	58
Reagents	107
Recipes	135
Relaxing insects	25
Reptiles	12
Revolving stages	101
Rooms	63
Safety cord	37
Schieck objectives	89
Scoop nets	19
Sea anemones	117
Sea urchins	117
Section cutters	127, 130
Section cutting	122
Section knife	128
Section lifter	129
Seiler's microtome	132
Serum	138
Setting insects	24
Shelf brackets	70
Shot	2
Silver nitrate	137
Simon's arsenical soap	141
Skeletons	16
Skimming	39
Skimming net	40
Skinning birds	9
Skinning mammals	4
Skulls	17
Sledge microtome	131
Snails	118
Soap for imbedding	127
Softening tissues	125
Solid eye-pieces	87
Sow bugs	34
Specific names	46
Spiders	33
Spiders, mounting	30
Spreading insects	24
Spreading board	24
Sponges	117

	PAGE
Stands for birds	10
Starfish	117
Sterling microtome	130
Storage jars	57
Storing specimens	57
Stretching paper	72
Substitutes for cork	72
Sugar for moths	141
Sunken net	39
Surface collecting	39
Swainson's soap	140
Tables	105
Table cases	71
Tablets	53
Tangle	38
Teasing tissues	123
Tightening cases	70
Tolles' instruments	95
Transparent soap	127
Transporting insects	26
Trawl	37
Triplets	82
Turtles	13
Typical collections	58
Use of microscope	96
Useful hints	135
Vertebrates	2, 119
Vertical camera	100
Vials	55
Washing the collections	37
Wing trawl	38
Wooden tablets	53
Work tables	105
Works on collecting	1
Zeiss microscopes	94
Zeiss objectives	89
Zentmayer microscopes	95